VERY SHORT INTRODUCTIONS are for anyone wanting
a stimulating and accessible way in to a new subject. They are
written by experts and have been translated into more than
40 different languages. The series began in 1995 and now covers
a wide variety of topics in every discipline. The VSI library contains
nearly 400 volumes—a Very Short Introduction to everything
from Indian philosophy to psychology and American history—and
continues to grow in every subject area.

Very Short Introductions available now:

ACCOUNTING Christopher Nobes
ADVERTISING Winston Fletcher
AFRICAN HISTORY John Parker and
    Richard Rathbone
AGNOSTICISM Robin Le Poidevin
ALEXANDER THE GREAT
    Hugh Bowden
AMERICAN HISTORY Paul S. Boyer
AMERICAN IMMIGRATION
    David A. Gerber
AMERICAN POLITICAL PARTIES
    AND ELECTIONS L. Sandy Maisel
AMERICAN POLITICS Richard M. Valelly
THE AMERICAN PRESIDENCY
    Charles O. Jones
ANAESTHESIA Aidan O'Donnell
ANARCHISM Colin Ward
ANCIENT EGYPT Ian Shaw
ANCIENT GREECE Paul Cartledge
THE ANCIENT NEAR EAST
    Amanda H. Podany
ANCIENT PHILOSOPHY Julia Annas
ANCIENT WARFARE Harry Sidebottom
ANGELS David Albert Jones
ANGLICANISM Mark Chapman
THE ANGLO-SAXON AGE John Blair
THE ANIMAL KINGDOM Peter Holland
ANIMAL RIGHTS David DeGrazia
THE ANTARCTIC Klaus Dodds
ANTISEMITISM Steven Beller
ANXIETY Daniel Freeman and
    Jason Freeman
THE APOCRYPHAL GOSPELS
    Paul Foster
ARCHAEOLOGY Paul Bahn

ARCHITECTURE Andrew Ballantyne
ARISTOCRACY William Doyle
ARISTOTLE Jonathan Barnes
ART HISTORY Dana Arnold
ART THEORY Cynthia Freeland
ASTROBIOLOGY David C. Catling
ATHEISM Julian Baggini
AUGUSTINE Henry Chadwick
AUSTRALIA Kenneth Morgan
AUTISM Uta Frith
THE AVANT GARDE David Cottington
THE AZTECS David Carrasco
BACTERIA Sebastian G. B. Amyes
BARTHES Jonathan Culler
THE BEATS David Sterritt
BEAUTY Roger Scruton
BESTSELLERS John Sutherland
THE BIBLE John Riches
BIBLICAL ARCHAEOLOGY Eric H. Cline
BIOGRAPHY Hermione Lee
THE BLUES Elijah Wald
THE BOOK OF MORMON Terryl Givens
BORDERS Alexander C. Diener and
    Joshua Hagen
THE BRAIN Michael O'Shea
THE BRITISH CONSTITUTION
    Martin Loughlin
THE BRITISH EMPIRE Ashley Jackson
BRITISH POLITICS Anthony Wright
BUDDHA Michael Carrithers
BUDDHISM Damien Keown
BUDDHIST ETHICS Damien Keown
CANCER Nicholas James
CAPITALISM James Fulcher
CATHOLICISM Gerald O'Collins

The Periodic Table: A Very Short Introduction

CAUSATION Stephen Mumford and
    Rani Lill Anjum
THE CELL Terence Allen and
    Graham Cowling
THE CELTS Barry Cunliffe
CHAOS Leonard Smith
CHILDREN'S LITERATURE
    Kimberley Reynolds
CHINESE LITERATURE Sabina Knight
CHOICE THEORY Michael Allingham
CHRISTIAN ART Beth Williamson
CHRISTIAN ETHICS D. Stephen Long
CHRISTIANITY Linda Woodhead
CITIZENSHIP Richard Bellamy
CIVIL ENGINEERING David Muir Wood
CLASSICAL LITERATURE William Allan
CLASSICAL MYTHOLOGY Helen Morales
CLASSICS Mary Beard and John Henderson
CLAUSEWITZ Michael Howard
CLIMATE Mark Maslin
THE COLD WAR Robert McMahon
COLONIAL AMERICA Alan Taylor
COLONIAL LATIN AMERICAN
    LITERATURE Rolena Adorno
COMEDY Matthew Bevis
COMMUNISM Leslie Holmes
COMPLEXITY John H. Holland
THE COMPUTER Darrel Ince
THE CONQUISTADORS Matthew
    Restall and Felipe Fernández-Armesto
CONSCIENCE Paul Strohm
CONSCIOUSNESS Susan Blackmore
CONTEMPORARY ART Julian Stallabrass
CONTEMPORARY FICTION
    Robert Eaglestone
CONTINENTAL PHILOSOPHY
    Simon Critchley
CORAL REEFS Charles Sheppard
COSMOLOGY Peter Coles
CRITICAL THEORY Stephen Eric Bronner
THE CRUSADES Christopher Tyerman
CRYPTOGRAPHY Fred Piper and
    Sean Murphy
THE CULTURAL
    REVOLUTION Richard Curt Kraus
DADA AND SURREALISM
    David Hopkins
DARWIN Jonathan Howard
THE DEAD SEA SCROLLS Timothy Lim
DEMOCRACY Bernard Crick
DERRIDA Simon Glendinning

DESCARTES Tom Sorell
DESERTS Nick Middleton
DESIGN John Heskett
DEVELOPMENTAL BIOLOGY
    Lewis Wolpert
THE DEVIL Darren Oldridge
DIASPORA Kevin Kenny
DICTIONARIES Lynda Mugglestone
DINOSAURS David Norman
DIPLOMACY Joseph M. Siracusa
DOCUMENTARY FILM
    Patricia Aufderheide
DREAMING J. Allan Hobson
DRUGS Leslie Iversen
DRUIDS Barry Cunliffe
EARLY MUSIC Thomas Forrest Kelly
THE EARTH Martin Redfern
ECONOMICS Partha Dasgupta
EDUCATION Gary Thomas
EGYPTIAN MYTH Geraldine Pinch
EIGHTEENTH-CENTURY
    BRITAIN Paul Langford
THE ELEMENTS Philip Ball
EMOTION Dylan Evans
EMPIRE Stephen Howe
ENGELS Terrell Carver
ENGINEERING David Blockley
ENGLISH LITERATURE Jonathan Bate
ENVIRONMENTAL ECONOMICS
    Stephen Smith
EPIDEMIOLOGY Rodolfo Saracci
ETHICS Simon Blackburn
THE EUROPEAN UNION John Pinder
    and Simon Usherwood
EVOLUTION Brian and
    Deborah Charlesworth
EXISTENTIALISM Thomas Flynn
THE EYE Michael Land
FAMILY LAW Jonathan Herring
FASCISM Kevin Passmore
FASHION Rebecca Arnold
FEMINISM Margaret Walters
FILM Michael Wood
FILM MUSIC Kathryn Kalinak
THE FIRST WORLD WAR
    Michael Howard
FOLK MUSIC Mark Slobin
FOOD John Krebs
FORENSIC PSYCHOLOGY David Canter
FORENSIC SCIENCE Jim Fraser
FOSSILS Keith Thomson

FOUCAULT Gary Gutting
FRACTALS Kenneth Falconer
FREE SPEECH Nigel Warburton
FREE WILL Thomas Pink
FRENCH LITERATURE John D. Lyons
THE FRENCH REVOLUTION
    William Doyle
FREUD Anthony Storr
FUNDAMENTALISM Malise Ruthven
GALAXIES John Gribbin
GALILEO Stillman Drake
GAME THEORY Ken Binmore
GANDHI Bhikhu Parekh
GENIUS Andrew Robinson
GEOGRAPHY John Matthews and
    David Herbert
GEOPOLITICS Klaus Dodds
GERMAN LITERATURE Nicholas Boyle
GERMAN PHILOSOPHY
    Andrew Bowie
GLOBAL CATASTROPHES Bill McGuire
GLOBAL ECONOMIC HISTORY
    Robert C. Allen
GLOBAL WARMING Mark Maslin
GLOBALIZATION Manfred Steger
THE GOTHIC Nick Groom
GOVERNANCE Mark Bevir
THE GREAT DEPRESSION AND THE
    NEW DEAL Eric Rauchway
HABERMAS James Gordon Finlayson
HAPPINESS Daniel M. Haybron
HEGEL Peter Singer
HEIDEGGER Michael Inwood
HERODOTUS Jennifer T. Roberts
HIEROGLYPHS Penelope Wilson
HINDUISM Kim Knott
HISTORY John H. Arnold
THE HISTORY OF ASTRONOMY
    Michael Hoskin
THE HISTORY OF LIFE
    Michael Benton
THE HISTORY OF
    MATHEMATICS Jacqueline Stedall
THE HISTORY OF MEDICINE
    William Bynum
THE HISTORY OF TIME
    Leofranc Holford-Strevens
HIV/AIDS Alan Whiteside
HOBBES Richard Tuck
HORMONES Martin Luck
HUMAN EVOLUTION Bernard Wood
HUMAN RIGHTS Andrew Clapham
HUMANISM Stephen Law
HUME A. J. Ayer
HUMOUR Noël Carroll
THE ICE AGE Jamie Woodward
IDEOLOGY Michael Freeden
INDIAN PHILOSOPHY Sue Hamilton
INFORMATION Luciano Floridi
INNOVATION Mark Dodgson and
    David Gann
INTELLIGENCE Ian J. Deary
INTERNATIONAL
    MIGRATION Khalid Koser
INTERNATIONAL RELATIONS
    Paul Wilkinson
INTERNATIONAL
    SECURITY Christopher S. Browning
ISLAM Malise Ruthven
ISLAMIC HISTORY Adam Silverstein
ITALIAN LITERATURE Peter
    Hainsworth and David Robey
JESUS Richard Bauckham
JOURNALISM Ian Hargreaves
JUDAISM Norman Solomon
JUNG Anthony Stevens
KABBALAH Joseph Dan
KAFKA Ritchie Robertson
KANT Roger Scruton
KEYNES Robert Skidelsky
KIERKEGAARD Patrick Gardiner
THE KORAN Michael Cook
LANDSCAPE ARCHITECTURE
    Ian H. Thompson
LANDSCAPES AND
    GEOMORPHOLOGY
    Andrew Goudie and Heather Viles
LANGUAGES Stephen R. Anderson
LATE ANTIQUITY Gillian Clark
LAW Raymond Wacks
THE LAWS OF THERMODYNAMICS
    Peter Atkins
LEADERSHIP Keith Grint
LINCOLN Allen C. Guelzo
LINGUISTICS Peter Matthews
LITERARY THEORY Jonathan Culler
LOCKE John Dunn
LOGIC Graham Priest
MACHIAVELLI Quentin Skinner
MADNESS Andrew Scull
MAGIC Owen Davies
MAGNA CARTA Nicholas Vincent

MAGNETISM Stephen Blundell
MALTHUS Donald Winch
MANAGEMENT John Hendry
MAO Delia Davin
MARINE BIOLOGY Philip V. Mladenov
THE MARQUIS DE SADE John Phillips
MARTIN LUTHER Scott H. Hendrix
MARTYRDOM Jolyon Mitchell
MARX Peter Singer
MATHEMATICS Timothy Gowers
THE MEANING OF LIFE Terry Eagleton
MEDICAL ETHICS Tony Hope
MEDICAL LAW Charles Foster
MEDIEVAL BRITAIN John
    Gillingham and Ralph A. Griffiths
MEMORY Jonathan K. Foster
METAPHYSICS Stephen Mumford
MICHAEL FARADAY Frank A.J.L. James
MICROECONOMICS Avinash Dixit
MODERN ART David Cottington
MODERN CHINA Rana Mitter
MODERN FRANCE Vanessa R. Schwartz
MODERN IRELAND Senia Pašeta
MODERN JAPAN Christopher Goto-Jones
MODERN LATIN AMERICAN
    LITERATURE
    Roberto González Echevarría
MODERN WAR Richard English
MODERNISM Christopher Butler
MOLECULES Philip Ball
THE MONGOLS Morris Rossabi
MORMONISM Richard Lyman Bushman
MUHAMMAD Jonathan A.C. Brown
MULTICULTURALISM Ali Rattansi
MUSIC Nicholas Cook
MYTH Robert A. Segal
THE NAPOLEONIC WARS
    Mike Rapport
NATIONALISM Steven Grosby
NELSON MANDELA Elleke Boehmer
NEOLIBERALISM Manfred Steger and
    Ravi Roy
NETWORKS Guido Caldarelli
    and Michele Catanzaro
THE NEW TESTAMENT
    Luke Timothy Johnson
THE NEW TESTAMENT AS
    LITERATURE Kyle Keefer
NEWTON Robert Iliffe
NIETZSCHE Michael Tanner

NINETEENTH-CENTURY
    BRITAIN Christopher Harvie and
    H. C. G. Matthew
THE NORMAN CONQUEST
    George Garnett
NORTH AMERICAN
    INDIANS Theda Perdue
    and Michael D. Green
NORTHERN IRELAND
    Marc Mulholland
NOTHING Frank Close
NUCLEAR POWER Maxwell Irvine
NUCLEAR WEAPONS
    Joseph M. Siracusa
NUMBERS Peter M. Higgins
NUTRITION David A. Bender
OBJECTIVITY Stephen Gaukroger
THE OLD TESTAMENT
    Michael D. Coogan
THE ORCHESTRA D. Kern Holoman
ORGANIZATIONS Mary Jo Hatch
PAGANISM Owen Davies
THE PALESTINIAN-ISRAELI CONFLICT
    Martin Bunton
PARTICLE PHYSICS Frank Close
PAUL E. P. Sanders
PENTECOSTALISM William K. Kay
THE PERIODIC TABLE Eric R. Scerri
PHILOSOPHY Edward Craig
PHILOSOPHY OF LAW Raymond Wacks
PHILOSOPHY OF SCIENCE
    Samir Okasha
PHOTOGRAPHY Steve Edwards
PLAGUE Paul Slack
PLANETS David A. Rothery
PLANTS Timothy Walker
PLATO Julia Annas
POLITICAL PHILOSOPHY David Miller
POLITICS Kenneth Minogue
POSTCOLONIALISM Robert Young
POSTMODERNISM Christopher Butler
POSTSTRUCTURALISM
    Catherine Belsey
PREHISTORY Chris Gosden
PRESOCRATIC PHILOSOPHY
    Catherine Osborne
PRIVACY Raymond Wacks
PROBABILITY John Haigh
PROGRESSIVISM Walter Nugent
PROTESTANTISM Mark A. Noll

PSYCHIATRY Tom Burns
PSYCHOLOGY Gillian Butler
    and Freda McManus
PURITANISM Francis J. Bremer
THE QUAKERS Pink Dandelion
QUANTUM THEORY John Polkinghorne
RACISM Ali Rattansi
RADIOACTIVITY Claudio Tuniz
RASTAFARI Ennis B. Edmonds
THE REAGAN REVOLUTION Gil Troy
REALITY Jan Westerhoff
THE REFORMATION Peter Marshall
RELATIVITY Russell Stannard
RELIGION IN AMERICA Timothy Beal
THE RENAISSANCE Jerry Brotton
RENAISSANCE ART
    Geraldine A. Johnson
REVOLUTIONS Jack A. Goldstone
RHETORIC Richard Toye
RISK Baruch Fischhoff and John Kadvany
RIVERS Nick Middleton
ROBOTICS Alan Winfield
ROMAN BRITAIN Peter Salway
THE ROMAN EMPIRE Christopher Kelly
THE ROMAN REPUBLIC
    David M. Gwynn
ROMANTICISM Michael Ferber
ROUSSEAU Robert Wokler
RUSSELL A. C. Grayling
RUSSIAN HISTORY Geoffrey Hosking
RUSSIAN LITERATURE Catriona Kelly
THE RUSSIAN REVOLUTION
    S. A. Smith
SCHIZOPHRENIA Chris Frith and
    Eve Johnstone
SCHOPENHAUER
    Christopher Janaway
SCIENCE AND RELIGION
    Thomas Dixon
SCIENCE FICTION David Seed
THE SCIENTIFIC REVOLUTION
    Lawrence M. Principe
SCOTLAND Rab Houston
SEXUALITY Véronique Mottier
SHAKESPEARE Germaine Greer
SIKHISM Eleanor Nesbitt
THE SILK ROAD James A. Millward
SLEEP Steven W. Lockley and
    Russell G. Foster

SOCIAL AND CULTURAL
    ANTHROPOLOGY
    John Monaghan and Peter Just
SOCIALISM Michael Newman
SOCIOLINGUISTICS John Edwards
SOCIOLOGY Steve Bruce
SOCRATES C. C. W. Taylor
THE SOVIET UNION Stephen Lovell
THE SPANISH CIVIL WAR
    Helen Graham
SPANISH LITERATURE Jo Labanyi
SPINOZA Roger Scruton
SPIRITUALITY Philip Sheldrake
STARS Andrew King
STATISTICS David J. Hand
STEM CELLS Jonathan Slack
STUART BRITAIN John Morrill
SUPERCONDUCTIVITY Stephen Blundell
SYMMETRY Ian Stewart
TEETH Peter S. Ungar
TERRORISM Charles Townshend
THEOLOGY David F. Ford
THOMAS AQUINAS Fergus Kerr
THOUGHT Tim Bayne
TIBETAN BUDDHISM
    Matthew T. Kapstein
TOCQUEVILLE Harvey C. Mansfield
TRAGEDY Adrian Poole
THE TROJAN WAR Eric H. Cline
TRUST Katherine Hawley
THE TUDORS John Guy
TWENTIETH-CENTURY
    BRITAIN Kenneth O. Morgan
THE UNITED NATIONS
    Jussi M. Hanhimäki
THE U.S. CONGRESS Donald A. Ritchie
THE U.S. SUPREME COURT
    Linda Greenhouse
UTOPIANISM Lyman Tower Sargent
THE VIKINGS Julian Richards
VIRUSES Dorothy H. Crawford
WITCHCRAFT Malcolm Gaskill
WITTGENSTEIN A. C. Grayling
WORK Stephen Fineman
WORLD MUSIC Philip Bohlman
THE WORLD TRADE
    ORGANIZATION Amrita Narlikar
WRITING AND SCRIPT
    Andrew Robinson

Eric R. Scerri

THE PERIODIC TABLE
A Very Short Introduction

OXFORD
UNIVERSITY PRESS

OXFORD

UNIVERSITY PRESS

Great Clarendon Street, Oxford OX2 6DP

Oxford University Press is a department of the University of Oxford.
It furthers the University's objective of excellence in research, scholarship,
and education by publishing worldwide in

Oxford New York

Auckland Cape Town Dar es Salaam Hong Kong Karachi
Kuala Lumpur Madrid Melbourne Mexico City Nairobi
New Delhi Shanghai Taipei Toronto

With offices in

Argentina Austria Brazil Chile Czech Republic France Greece
Guatemala Hungary Italy Japan Poland Portugal Singapore
South Korea Switzerland Thailand Turkey Ukraine Vietnam

Oxford is a registered trade mark of Oxford University Press
in the UK and in certain other countries

Published in the United States
by Oxford University Press Inc., New York

British Library Cataloguing in Publication Data

Data available

Library of Congress Cataloging in Publication Data

Data available

Typeset by SPI Publisher Services, Pondicherry, India
Printed in Great Britain on acid-free paper by
Ashford Colour Press Ltd, Gosport, Hampshire

ISBN 978-0-19-958249-5

7 9 10 8 6

Contents

Acknowledgements xi

List of illustrations xiii

Preface xvii

1 The elements 1

2 A quick overview of the modern periodic table 10

3 Atomic weight, triads, and Prout 30

4 Steps towards the periodic table 42

5 The Russian genius – Mendeleev 58

6 Physics invades the periodic table 72

7 Electronic structure 84

8 Quantum mechanics 98

9 Modern alchemy: from missing elements
   to synthetic elements 109

10 Forms of the periodic table 122

Further reading 139

Index 141

Acknowledgements

I acknowledge the help of all editors and staff at OUP, including Jeremy Lewis for suggesting the idea of a VSI. I also thank colleagues, students and librarians who helped me during the preparation of this book. Last but not least, I thank my wife Elisa for her love and patience.

List of illustrations

1 The Platonic solids **2**

From O. Benfey, 'Precursors and Cocursors of the Mendeleev Table: The Pythagorean Spirit in Element Classification', *Bulletin for the History of Chemistry*, 13–14, 60–66, 1992–1993, figure on p. 60. By permission from the publisher

2 Lavoisier's list of elements as simple substances **4**

From A-L. Lavoisier, *Traite Elementaire de Chimie*, Cuchet, Paris, 1789, p. 192

3 Names and symbols of ancient elements **8**

From V. Ringnes, 'Origin of the Names of Chemical Elements', from Journal of Chemical Education, 66, 731–738, 1989, p. 731. By permission, © American Chemical Society

4 Medium-long-form table **13**

5 Short-form periodic table published by Mendeleev in 1871 **15**

From D. I. Mendeleev, 'Sootnoshenie svoistv s atomnym vesom elementov', *Zhurnal Russkeo Fiziko-Khimicheskoe Obshcestv*, 1, 60–77, 1869, table on p. 70

6 Long-form periodic table **18**

7 Periodic tables before and after Seaborg's modification **25**

8 Richter's table of equivalent weights as modified by Fischer in 1802 **32**

From E. G. Fischer, *Claude Louis Bethollet uber die Gsetze der Verwandtschaft*, Berlin, 1802. Table on p. 229

9 The 20 triads of Lenssen **38**

From E. Lenssen, 'Uber die gruppirung der elemente nach ihrem chemisch-physikalishen characckter', *Annalen der Chemie und Pharmazie*, 103, 121–131, 1857

10 Gmelin's table of elements **39**

From L. Gmelin, *Handbuch der anorganischen chemie* 4th ed, Heidelberg, 1843, vol 1, p. 52

11 Kremers' atomic weight differences for the oxygen series **39**

12 De Chancourtois' telluric screw **44**

From A.E. Beguyer De Chancourtois, 'Vis Tellurique: Classement naturel des corps simples ou radicaux, obtenu au moyen d'un systeme de classification helicoidale et numerique', *Comptes Rendus de L'Academie*, 54, 757-761, 840-843, 967-971, 1862. Redrawn in J. van Spronsen, The Periodic System of the Chemical Elements, the First One Hundred Years, Elsevier, Amsterdam, 1969, p. 99. By permission from the publisher

13 First seven of Newlands' grouping of elements in 1863 **46**

From J. A. R. Newlands, 'On Relations among the Equivalents', *Chemical News*, 7, 70–72, 1863, table on p. 71

14 Newlands' table for the law of octaves, presented to the Chemical Society in 1866 **48**

From J. A. R. Newlands, as reported in *Proceedings of Chemical Societies: Chemical Society, Chemical News*, 13 113–114, 1866, table on p.113

15 Odling's third table of differences **49**

From W. Odling, 'On the Proportional Numbers of the Elements', *Quarterly Journal of Science*, 1, 642–648, 1864, table on p. 645

16 Hinrichs' table of planetary distances, 1864 **51**

From G. D. Hinrichs, 'The Density, Rotation and Relative Age of the Planets', *American Journal of Science and Arts*, 2(37), 36–56, 1864, table on p. 43

17 Hinrichs' periodic system **52**

From G. D. Hinrich, Programm der Atomechanik oder die Chemie eine Mechanik de Pantome, Augustus Hageboek, Iowa City, IA, 1867. As simplified by J. van Spronsen, The Periodic System of the Chemical Elements, the First One Hundred Years, Elsevier, Amsterdam, 1969. By permission from the publisher

18 Lothar Meyer's periodic system of 1862 **54**

From J. Lothar Meyer, *Die modern theorien und ihre Bedeutung fur die chemische Statisik*, Breslau (Wroclaw), 1864, p. 135

19 Lothar Meyer's periodic system of 1868 **56**

From J. Lothar Meyer, unpublished system of 1868

20 The predicted and observed properties of eka-silicon (germanium) **67**

21 Mendeleev's predictions, successful and otherwise **68**

22 Marie Curie **73**

AIP Emilio Segrè Visual Archives, W.F. Meggers Gallery of Nobel Laureates

23 Bohr's original 1913 scheme for electronic configurations of atoms **87**

From N. Bohr, 'On the Constitution of Atoms and Molecules', *Philosophical Magazine*, 26, 476–502, 1913, 497

24 Bohr's 1923 electronic configurations based on two quantum numbers **88**

From N. Bohr, 'Linienspektren und Atombau', *Annalen der Physik*, 71, 228–288, 1923, p. 260.

25 Combination of four quantum numbers to explain the total number of electrons in each shell **90**

26 Order of filling of atomic orbitals **91**

27 Lewis's sketch of atomic cubes **93**

From G. N. Lewis, unpublished memorandum of March 28, 1902

28 Lewis's depiction of a double bond between two atoms **94**

From A. Stranges, Electrons and Valence, Texas A&M University Press, College Station, TX, 1982, p.213. By permission from the publisher

29 Lewis's outer electronic structures for 29 elements **95**

30 Langmuir's periodic table **96**

From W. Langmuir, 'Arrangement of Electrons in Atoms and Molecules', *Journal of the American Chemical Society*, 41, 868–934, 1919, p. 874

31 Niels Bohr **101**

AIP Emilio Segrè Visual Archives, Weber Collection

32 Theoretically calculated and observed ionization energies of elements 1 to 53 **103**

From E. Clementi, Computational Aspects of Large Chemical Systems, Lecture Notes in Chemistry, vol 19, Springer-Verlag, Berlin, 1980, p.12. By permission from the publisher.

33 Boundary conditions and the emergence of quantization **107**

From R. Chang, Physical Chemistry for the Chemical Biological Sciences, University Science Books, Sausalito, CA, 2000, p. 576. By permission from the publisher

34 s, p, d, and f orbitals **107**

35 Fragment of periodic table showing groups 3 to 12 inclusive **119**

36 Sublimation enthalpies for elements in group 7, showing that element 107 is a genuine member of this group **120**

37 Benfey's periodic system **123**

38 Dufour's periodic tree **124**

39 The left-step periodic table by Charles Janet **126**

40 Periodic table based on maximizing atomic number triads **133**

41 Long-form periodic tables showing alternative placements of Lu and Lr **136**

42 Third option for presenting long-form periodic table **137**

Preface

Much has been written about the wonders of the periodic table. Here are just a few examples.

> The Periodic Table is nature's Rosetta stone. To the uninitiated, it's just 100-plus numbered boxes, each containing one or two letters, arranged with an odd, skewed symmetry. To chemists, however, the periodic table reveals the organizing principles of matter, which is to say, the organizing principles of chemistry. At a fundamental level, all of chemistry is contained in the periodic table.
>
> That's not to say, of course, that all of chemistry is obvious from the periodic table. Far from it. But the structure of the table reflects the electronic structure of the elements, and hence their chemical properties and behavior. Perhaps it would be more appropriate to say that all of chemistry starts with the periodic table.
>
> (Rudy Baum, *C&EN Special Issue on Elements*)

Astronomer Harlow Shapley writes:

> The periodic Table is probably the most compact and meaningful compilation of knowledge that man has yet devised. The periodic table does for matter what the geological age table does for cosmic

time. Its history is the story of man's great conquests in the microcosmos.

As Robert Hicks, a historian of chemistry, says on an Internet podcast:

> Perhaps the most recognizable icon in all of science is the periodic table of elements. This chart has become our model for how atoms and molecules arrange themselves to create matter as we know it. How the world is organized on the most minute level. Throughout history the periodic table has changed. Newly discovered elements have been added to it and other elements have been disproved and either modified or removed. In this way the periodic table acts like a store of the history of chemistry, a template for current developments and a basis for the future of the chemical sciences...a map of the world's most basic building blocks.

As a final example for now, here is C. P. Snow, a physical chemist known for his writings on the 'two cultures':

> [On learning about the table] For the first time I saw a medley of haphazard facts fall into line and order. All the jumbles and recipes and hotchpotch of the inorganic chemistry of my boyhood seemed to fit themselves into the scheme before my eyes — as though one were standing beside a jungle and it suddenly transformed itself into a Dutch garden.

What is remarkable about the periodic table is its simultaneous simplicity and familiarity coupled with its truly fundamental status in science. The simplicity is alluded to in the above quotations. The periodic table seems to organize the fundamental components of all matter. It is also familiar to most people. Almost everybody with even just an elementary acquaintance with chemistry can recall the existence of the periodic table even after everything else they ever learned in chemistry may have been forgotten. The periodic table is almost as familiar as the chemical formula for water. It has become a true cultural icon that is used by artists, advertisers, and of course scientists of all kinds.

At the same time, the periodic table is more than just a device for teaching and learning chemistry. It reflects the natural order of things in the world and, as far as we know, in the whole universe. It consists of groupings of elements into vertical columns. If a chemist or even a student of chemistry knows the properties of one typical element in any group, such as sodium, he or she will have a good idea of the properties of elements in the same group, such as potassium, rubidium, and caesium.

More fundamentally, the order inherent in the periodic table has led to a deep knowledge of the structure of the atom and the notion that electrons essentially circle the nucleus in specific shells and orbitals. These arrangements of electrons in turn serve to rationalize the periodic table. They explain why, broadly speaking, the elements sodium, potassium, rubidium, and so on fall into the same group in the first place. But more importantly, the understanding of atomic structure that was first arrived at in trying to understand the periodic table has been applied in many other fields of science. This knowledge has contributed to the development of, first, the old quantum theory and then to its more mature cousin, quantum mechanics, a body of knowledge that continues to be the fundamental theory of physics that can explain the behaviour not just of all matter, but also all forms of radiation such as visible light, X-rays, and ultraviolet light.

Unlike most scientific discoveries made in the 19th century, the periodic table has not been refuted by discoveries made in the 20th and 21st centuries. Rather, discoveries in modern physics, in particular, have served to refine the periodic table and to tidy up some remaining anomalies. But its overall form and validity have remained intact as another testament to the power and depth of this system of knowledge.

Before examining the periodic table, we will consider its occupants, the elements. Then we will take a quick look at the modern periodic table and some of its variants, before looking into its history and how we got to our present level of understanding, from Chapter 3.

Chapter 1
The elements

The ancient Greek philosophers recognized just four elements, earth, water, air, and fire, all of which survive in the astrological subdivision of the twelve signs of the zodiac. Some of these philosophers believed that these different elements consisted of microscopic components with differing shapes, and that this explained the various properties of the elements. The basic shapes of the four elements were thought to be those of the Platonic solids (Figure 1) made up entirely of the same two-dimensional shapes like triangles or squares. The Greeks believed that earth consisted of microscopic cubic particles. This association was made because, of all the Platonic solids, the cube possesses the faces with the largest surface area. The liquidity of water was explained by an appeal to the smoother shape possessed by the icosahedron, while fire was said to be painful to the touch because it consisted of the sharp particles in the form of tetrahedra. Air was thought to consist of octahedra since that was the only remaining Platonic solid. Some time later, a fifth Platonic solid, the dodecahedron, was discovered by mathematicians, and this led Aristotle to propose that there might be a fifth element, or 'quintessence', which became known also as ether.

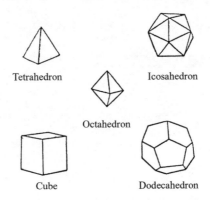

1. The Platonic solids, each one associated with one of the ancient elements

Today, the notion that elements are made up of Platonic solids is regarded as incorrect, but it was the origin of the fruitful notion that macroscopic properties of substances are governed by the structures of the microscopic components of which they are comprised. These 'elements' survived well into the Middle Ages and beyond, augmented with a few others discovered by the alchemists, the precursors of modern-day chemists. The best-known goal of the alchemists was to bring about the transmutation of elements. In particular, they attempted to change the base metal lead into the noble metal gold, whose colour, rarity, and chemical inertness has made it one of the most treasured substances since the dawn of civilization.

But in addition to being regarded as substances that could actually exist, Greek philosophers thought of the 'elements' as principles, or as tendencies and potentialities that gave rise to the observable properties of the elements. This rather subtle distinction between the abstract form of an element and its observable form has played an important role in the development of chemistry, although these days the more subtle meaning is not very well understood even by professional chemists. The notion of an abstract element has nonetheless served as a fundamental guiding principle to some of the pioneers of the periodic system such as Dimitri Mendeleev, its major discoverer.

According to most textbook accounts, chemistry only began properly when it turned its back on ancient Greek wisdom and alchemy and on this seemingly mystical understanding of the nature of elements. The triumph of modern science is generally regarded as resting on direct experimentation, which holds that only what can be observed should count. Not surprisingly, the more subtle and perhaps more fundamental sense of the concept of elements has generally been rejected. For example, Antoine Lavoisier took the view that an element should be defined by an appeal to empirical observation, thus relegating the role of abstract elements or elements as principles. Lavoisier held that an element should be defined as a material substance that has yet to be broken down into any more fundamental components. In 1789, Lavoisier published a list of 33 simple substances, or elements, according to this empirical criterion (Figure 2). The ancient elements of earth, water, air, and fire, which had by now been shown to consist of simpler substances, were correctly omitted from the list of elements, but so too was the notion of an abstract element.

Many of the substances in Lavoisier's list would qualify as elements by modern standards, while others like *lumière* (light) and *calorique* (heat) are no longer regarded as elements. Rapid advances in techniques of separation and characterization of chemical substances over the forthcoming years would help chemists expand this list. The important technique of spectroscopy, which measures the emission and absorption spectra of various kinds of radiation, would eventually yield a very accurate means by which each element could be identified through its 'fingerprint'. Today, we recognize about 90 naturally occurring elements. Moreover, an additional 25 or so elements have been artificially synthesized.

	Noms nouveaux.	Noms anciens correspondans.
Substances simples qui appartiennent aux trois règnes, & qu'on peut regarder comme les élémens des corps.	Lumière	Lumière.
	Calorique........	Chaleur.
		Principe de la chaleur.
		Fluide igné.
		Feu.
		Matière du feu & de la chaleur.
	Oxygène	Air déphlogistiqué.
		Air empiréal.
		Air vital.
		Base de l'air vital.
	Azote	Gaz phlogistiqué.
		Mofète.
		Base de la mofète.
	Hydrogène	Gaz inflammable.
		Base du gaz inflammable.
Substances simples non métalliques oxidables & acidifiables.	Soufre	Soufre.
	Phosphore	Phosphore.
	Carbone	Charbon pur.
	Radical muriatique .	Inconnu.
	Radical fluorique...	Inconnu.
	Radical boracique..	Inconnu.
Substances simples métalliques oxidables & acidifiables.	Antimoine	Antimoine.
	Argent	Argent.
	Arsenic	Arsenic.
	Bismuth	Bismuth.
	Cobalt	Cobalt.
	Cuivre..........	Cuivre.
	Etain	Etain.
	Fer............	Fer.
	Manganèse.......	Manganèse.
	Mercure	Mercure.
	Molybdène	Molybdène.
	Nickel..........	Nickel.
	Or	Or.
	Platine	Platine.
	Plomb	Plomb.
	Tungstène	Tungstène.
	Zinc	Zinc.
Substances simples salifiables terreuses.	Chaux..........	Terre calcaire, chaux.
	Magnésie	Magnésie, base du sel d'epsom.
	Baryte	Barote, terre pesante.
	Alumine	Argile, terre de l'alun, base de l'alun.
	Silice	Terre siliceuse, terre vitrifiable.

2. Lavoisier's list of elements as simple substances

The discovery of the elements

Some elements like iron, copper, gold, and silver have been known since the dawn of civilization. This reflects the fact that these elements can occur in uncombined form or are easy to separate from the minerals in which they occur.

Historians and archaeologists refer to certain epochs in human history as the Iron Age or the Bronze Age (bronze is an alloy of copper and tin). The alchemists added several more elements to the list, including sulphur, mercury, and phosphorus. In relatively modern times, the discovery of electricity enabled chemists to isolate many of the more reactive elements, which, unlike copper and iron, could not be obtained by heating their ores with charcoal (carbon). There have been a number of major episodes in the history of chemistry when half a dozen or so elements were discovered within a period of a few years. For example, the English chemist Humphry Davy made use of electricity, or more specifically the technique of electrolysis, to isolate about ten elements, including calcium, barium, magnesium, sodium, and chlorine.

Following the discovery of radioactivity and nuclear fission, yet more elements were discovered. The last seven elements to be isolated within the limits of the naturally occurring elements were protactinium, hafnium, rhenium, technetium, francium, astatine, and promethium, between the years 1917 and 1945. One of the last gaps to be filled was that corresponding to element 43, which became known as technetium from the Greek *techne*, meaning 'artificial'. It was 'manufactured' in the course of radio-chemical reactions that would not have been feasible before the advent of nuclear physics. Nevertheless, it now appears that technetium does occur naturally on earth, although in minuscule amounts.

Names of the elements

Part of the appeal of the periodic table derives from the individual nature of the elements such as their colours or how they feel to the touch. Much interest also lies in their names. The chemist and concentration camp survivor Primo Levi wrote a much-acclaimed book called simply *The Periodic Table* in which each chapter is named after an element. The book is mostly about his relations and acquaintances, but each anecdote is motivated by Levi's love of a particular element. The well-known neurologist and author Oliver Sacks has written a book called *Uncle Tungsten* in which he tells of his fascination with the elements, with chemistry, and in particular with the periodic table. Even more recently, two popular books on the elements have been written by Sam Kean and Hugh Aldersey-Williams. I think it is fair to say that the appeal of the elements in the public imagination has now truly arrived.

During the many centuries over which the elements have been discovered, many different approaches have been used to give them their names. Promethium, element 61, takes its name from Prometheus, the god who stole fire from heaven and gave it to human beings only to be punished for this act by Zeus. The connection of this tale to element 61 is in the heroic effort that was needed to isolate it, by analogy to the heroic and dangerous feat of Prometheus in Greek mythology. Promethium is one of the few elements that do not occur naturally on the earth. It was obtained as a decay product from the fission of another element, uranium.

Planets and other celestial bodies have also been used to name some elements. Helium is named after *helios*, the Greek name for the Sun. It was first observed in the spectrum of the Sun in 1868, and it was not until 1895 that it was first identified in terrestrial samples. Similarly, we have palladium after the asteroid Pallas, which in turn was named after Pallas, the Greek goddess of wisdom. The element cerium is named after Ceres, the first asteroid to be discovered, in the year 1801. Uranium is named after the planet Uranus, both the planet and the element having been discovered in the 1780s. In many of these cases too, the mythological theme persists. Uranus, for example, was the god of heaven in Greek mythology.

Many elements get their names from colours. The yellow-green gas chlorine comes from the Greek word *khloros*, which denotes the colour yellow-green. Caesium is named after the Latin *cesium*, which means grey-blue, because it has prominent grey-blue lines in its spectrum. The salts of the element rhodium often have a pink colour, and this explains why the name of the element was chosen from *rhodon*, the Greek for rose. The metal thallium gets its name from the Latin *thallus*, meaning green twig. It is an element that was discovered by the British chemist William Crookes from the prominent green line in its spectrum.

A large number of element names have come from the place where their discoverer lived, or wished to honour, such as americium, berkelium, californium, darmstadtium, europium, francium, germanium, hassium, polonium, gallium, hafnium (from *hafnia*, the Latin name for Copenhagen), lutetium (from *Lutetia*, Latin for Paris), rhenium (from the region of the Rhine river), ruthenium (from *Rus*, Latin for an area which includes present-day western Russia, Ukraine, Belarus, and parts of Slovakia and Poland). Yet other element names are derived from geographical locations connected with minerals in which they were found. This category includes the case of four elements named after the Swedish village of Ytterby, which lies close to Stockholm. Erbium, terbium, ytterbium, and yttrium were all found in ores located around this village, while a fifth element, holmium, was named after the Latin name for Stockholm.

In cases of more recently synthesized elements, their names come from those of the discoverer or a person whom the discoverers wished to honour. For example, we have bohrium, curium, einsteinium, fermium, lawrencium, meintnerium, mendeleevium, nobelium, roengenium, rutherfordium, and seaborgium.

The naming of the later transuranium elements has featured nationalistic controversies and, in some cases, bitter disputes over who first synthesized the element and who should therefore be given the honour of selecting a name for it. In an attempt to resolve such disputes, the International Union of Pure and Applied Chemistry (IUPAC) decreed that the elements should be named impartially and systematically with the Latin numerals for the atomic number of the element in each case. Element 105, for example, would be known as un-nil-pentium, while element 106 would be un-nil-hexium. But more recently, after much deliberation on some of these later superheavy elements, IUPAC has returned the naming rights to the discoverers or synthesizers who were judged to have established priority in each case. Elements 105 and 106 are now called dubnium and seaborgium respectively.

The symbols that are used to depict each element in the periodic table also have a rich and interesting story. In alchemical times, the symbols for the elements often coincided with those of the planets from which they were named, or with which they were associated (Figure 3). The element mercury, for example, shared

Metal						
gold	silver	iron	mercury	tin	copper	lead
Symbol						
○	☽	♂	☿	♃	♀	♄
Celestial Body						
Sun	Moon	Mars	Mercury	Jupiter	Venus	Saturn
Day						
Lat. *(dies)* Solis	Lunae	Martis	Mercurii	Jovis *(pater)*	Veneris	Saturni
Fr. dimanche	lundi	mardi	mercredi	jeudi	vendredi	samedi
Eng. Sunday	Monday	Tuesday	Wednesday	Thursday	Friday	Saturday

3. **Names and symbols of ancient elements**

the same symbol as that of Mercury the innermost planet in the solar system. Copper was associated with the planet Venus, and both the element and the planet shared the same symbol.

When John Dalton published his atomic theory in 1805, he retained several of the alchemical symbols for the elements. These were rather cumbersome, however, and did not lend themselves easily to reproduction in articles and books. The modern use of letter symbols was introduced by the Swedish chemist Jöns Jacob Berzelius in 1813.

A small minority of elements in the modern periodic table are represented by a single letter of the alphabet. These include hydrogen, carbon, oxygen, nitrogen, sulphur, and fluorine, which appear as H, C, O, N, S, and F. Most elements are depicted by two letters, the first of which is a capital letter and the second a lower-case letter. For example, we have Kr, Mg, Ne, Ba, Sc, for krypton, magnesium, neon, barium, and scandium, respectively. Some of the two-letter symbols are by no means intuitively obvious, such as Cu, Na, Fe, Pb, Hg, Ag, Au, etc., which are derived from the Latin names for the elements copper, sodium, iron, lead, mercury, silver, and gold. Tungsten is represented by a W after the German name for the element, which is wolfram.

Chapter 2
A quick overview of the modern periodic table

The modern periodic table

The manner in which the elements are arranged in rows and columns in the periodic table reveals many relationships among them. Some of these relationships are very well known while others may still await discovery. In the 1980s, scientists discovered that superconductivity, meaning the flow of an electric current with a zero resistance, occurred at much higher temperatures than had previously been observed. From typical values of 20K or less, the superconducting temperature very quickly leapt to values such as 100K. The discovery of these high-temperature superconductors came about when the elements lanthanum, copper, oxygen, and barium were combined together to form a complicated compound that happened to display high-temperature superconductivity. There followed a flurry of worldwide activity in an effort to raise the temperature at which the effect could be maintained. The ultimate goal was to achieve room-temperature superconductivity, which would allow technological breakthroughs such as levitating trains gliding effortlessly along superconducting rails. One of the main principles used in this quest was the periodic table of the elements. The table allowed researchers to replace some of the elements in the compound with others that are known to behave in a similar manner, in order to then examine the outcome on their superconducting behaviour. This is how the element yttrium was incorporated into a new set of superconducting compounds, to produce a superconducting temperature of 93K in the compound $YBa_2Cu_3O_7$. This knowledge, and undoubtedly much more, lies dormant within the periodic system waiting to be discovered and put to good use.

Even more recently, a new class of high-temperature superconductors has been discovered. They are the oxypnictides, a class of materials including oxygen, a pnictogen (group 15 elements) and one or more other elements. Interest in these compounds increased dramatically after the publication of the superconducting properties of LaOFeP and LaOFeAs, which were discovered in 2006 and 2008 respectively. Once again, the idea of using arsenic (As), as in the latter compound, came from its position, which is directly below phosphorus in the periodic table.

Chemical analogies between elements in the same group are also of great interest in the field of medicine. For example, the element beryllium sits at the top of group 2 of the periodic table and above magnesium. Because of the similarity between these two elements, beryllium can replace the element magnesium that is essential to human beings. This behaviour accounts for one of the many ways in which beryllium is toxic to humans. Similarly, the element cadmium lies directly below zinc in the periodic table, with the result that cadmium can replace zinc in many vital enzymes. Similarities can also occur between elements lying in adjacent positions in rows of the periodic table. For example, platinum lies next to gold. It has long been known that an inorganic compound of platinum called cis-platin can cure various forms of cancer. As a result, many drugs have been developed in which gold atoms are made to take the place of platinum, and this has produced some successful new drugs.

One final example of the medical consequences of how elements are placed in the periodic table is provided by rubidium, which lies directly below potassium in group 1 of the table. As in the previous cases mentioned, atoms of rubidium can mimic those of potassium, and so like potassium can easily be absorbed into the human body. This behaviour is exploited in monitoring techniques, since rubidium is attracted to cancers, especially those occurring in the brain.

The conventional periodic table consists of rows and columns. Trends can be observed among the elements going across and down the table. Each horizontal row represents a single period of the table. On crossing a period, one passes from metals such as potassium and calcium on the left, through transition metals such as iron, cobalt, and nickel, then through some semi-metallic elements like germanium, and on to some non-metals such as arsenic, selenium, and bromine, on the right side of the table. In general, there is a smooth gradation in chemical and physical properties as a period is crossed, but exceptions to this general rule abound and make the study of chemistry a fascinating and unpredictably complex field.

Metals themselves can vary from soft dull solids like sodium or potassium to hard shiny substances like chromium, platinum, or iron. Non-metals, on the other hand, tend to be solids or gases, such as carbon and oxygen respectively. In terms of their appearance, it is sometimes difficult to distinguish between solid metals and solid non-metals. To the layperson, a hard and shiny non-metal may seem to be more metallic than a soft metal like sodium. The periodic trend from metals to non-metals is repeated with each period, so that when the rows are stacked, they form columns, or groups, of similar elements. Elements within a single group tend to share many important physical and chemical properties, although there are many exceptions.

Recently, the International Union of Pure and Applied Chemistry (IUPAC) recommended that groups should be sequentially numbered with an Arabic numeral from left to right, as groups 1

1	2	3	4	5	6	7	8	9	10	11	12	13	14	15	16	17	18
H																	He
Li	Be											B	C	N	O	F	Ne
Na	Mg											Al	Si	P	S	Cl	Ar
K	Ca	Sc	Ti	V	Cr	Mn	Fe	Co	Ni	Cu	Zn	Ga	Ge	As	Se	Br	Kr
Rb	Sr	Y	Zr	Nb	Mo	Tc	Ru	Rh	Pd	Ag	Cd	In	Sn	Sb	Te	I	Xe
Cs	Ba	Lu	Hf	Ta	W	Re	Os	Ir	Pt	Au	Hg	Tl	Pb	Bi	Po	At	Rn
Fr	Ra	Lr	Rf	Db	Sg	Bh	Hs	Mt	Ds	Rg	Cn						

La	Ce	Pr	Nd	Pm	Sm	Eu	Gd	Tb	Dy	Ho	Er	Tm	Yb
Ac	Th	Pa	U	Np	Pu	Am	Cm	Bk	Cf	Es	Fm	Md	No

4. Medium-long-form table

to 18, without the use of the letters A or B which can be seen on older periodic tables (Figure 4).

Forms of the periodic table

There have been quite literally over 1,000 periodic tables published in print, or more recently on the Internet. How are they all related? Is there one optimal periodic table? These are questions that will be explored in this book since they can teach us many interesting things about modern science.

One aspect of this question should be addressed right away. One of the ways of classifying the periodic tables that have been published is to consider three basic formats. First of all, there are the originally produced short-form tables published by the pioneers of the periodic table like Newlands, Lothar Meyer, and Mendeleev, all of which will be examined more closely in due course (Figure 5).

These tables essentially crammed all the then known elements into eight vertical columns or groups. They reflect the fact that, broadly speaking, the elements seem to recur after an interval of eight elements if the elements are arranged in a natural sequence (another topic to be discussed). As more information was gathered on the properties of the elements, and as more elements were discovered, a new kind of arrangement called the medium-long-form table (Figure 4) began to gain prominence. Today, this form is almost completely ubiquitous. One odd feature is that the main body of the table does not contain all the elements. If you look at Figure 4, you will find a break between elements 56 and 71, and then again between elements 88 and 103. The 'missing' elements are grouped together in what looks like a separate footnote that lies below the main table.

This act of separating off the rare earth elements, as they have traditionally been called, is performed purely for convenience. If it were not carried out, the periodic table would appear much wider, 32 elements wide to be precise, instead of 18 elements wide. The 32-wide element format does not lend itself readily to being reproduced on the inside cover of chemistry textbooks or on large wall-charts that hang in lecture rooms and laboratories. But if the elements are shown in this expanded form, as they sometimes are, one has the long-form periodic table, which may be said to be more correct than the familiar medium-long form, in the sense that the sequence of elements is unbroken (Figure 6).

But what are the occupants of the periodic table? Let us go back to the periodic table in general and let us choose the familiar medium-long format in order to put some flesh onto the skeleton, or framework, provided by this two-dimensional grid. How were the elements discovered? What are the elements like? How do they differ as we move down a column of the periodic table or across a horizontal period?

Typical groups of elements in the periodic table

On the extreme left of the table, group 1 contains such elements as the metals sodium, potassium, and rubidium. These are unusually soft and reactive substances, quite unlike what are normally considered as metals, such as iron, chromium, gold, or silver. The metals of group 1 are so reactive that merely placing a small piece of one of them into pure water gives rise to a vigorous reaction that produces hydrogen gas and leaves behind a colourless alkaline solution. The elements in group 2 include magnesium, calcium, and barium, and tend to be less reactive than those of group 1 in most respects.

Moving to the right, one encounters a central rectangular block of elements collectively known as the transition metals, which includes such examples as iron, copper, and zinc. In early periodic tables, known as short-form tables (Figure 5), these elements were placed among the groups of what are now called the main group elements.

MENDELÉEFF'S TABLE I.—1871.

Series	GROUP I. R_2O.	GROUP II. RO.	GROUP III. R_2O_3.	GROUP IV. RH_4. RO_2.	GROUP V. RH_3. R_2O_5.	GROUP VI. RH_2. RO_3.	GROUP VII. RH. R_2O_7.	GROUP VIII. RO_4.
1	H=1							
2	Li=7	Be=9.4	B=11	C=12	N=14	O=16	F=19	
3	Na=23	Mg=24	Al=27.3	Si=28	P=31	S=32	Cl=35.5	
4	K=39	Ca=40	—=44	Ti=48	V=51	Cr=52	Mn=55	Fe=56, Ce=59 Ni=59, Cu=63
5	(Cu=63)	Zn=65	—=68	—=72	As=75	Se=78	Br=80	
6	Rb=85	Sr=87	? Y=88	Zr=90	Nb=94	Mo=96	—=100	Ru=104, Rh=104 Pd=106, Ag=108
7	(Ag=108)	Cd=112	In=113	Sn=118	Sb=122	Te=125	I=127	
8	Cs=133	Ba=137	? Di=138	? Ce=140	....	....	....	
9	....	....	....	....	....	...	....	
10	....	....	? Er=178	? La=180	Ta=182	W=184	....	Os=195, In=197 Pt=198, Au=199
11	(Au=199)	Hg=200	Tl=204	Pb=207	Bi=208	....	....	
12	....	....	....	Th=231	....	U=240	....	....

5. Short-form periodic table, published by Mendeleev in 1871

Several valuable features of the chemistry of these elements are lost in the modern table because of the manner in which they have been separated from the main body of the table, although the advantages of this later organization outweigh these losses. To the right of the transition metals, in the medium-long-form table, lies another block of representative elements starting with group 13 and ending with group 18, the noble gases on the extreme right of the table.

Sometimes the properties a group shares are not immediately obvious. This is the case with group 14, which consists of carbon, silicon, germanium, tin, and lead. Here, one notices a great diversity on progressing down the group. Carbon, at the head of the group, is a non-metal solid that occurs in three completely different structural forms (diamond, graphite, and fullerenes) and which forms the basis of all living systems. The next element below, silicon, is a semi-metal which, interestingly, forms the basis of artificial life, or at least artificial intelligence, since it lies at the heart of all computers. The next element down, germanium, is a more recently discovered semi-metal that was predicted by Mendeleev and later found to have many of the properties he foresaw. On moving down to tin and lead, one arrives at two metals known since antiquity. In spite of this wide variation among them, in terms of metal–non-metal behaviour, the elements of group 14 nevertheless are similar in an important chemical sense in that they all display a maximum combining power, or valency, of four.

The apparent diversity of the elements in group 17 is even more pronounced. The elements fluorine and chlorine, which head the group, are both poisonous gases. The next member, bromine, is one of the only two known elements that exist as liquids at room temperature, the other one being the metal mercury. Moving further down the group, one then encounters iodine, a violet-black solid element. If a novice chemist were asked to group these elements according to their appearances, it is unlikely that he or she would consider classifying together fluorine, chlorine, bromine, and iodine. This is one instance where the subtle distinction between the observable and the abstract senses of the concept of an element can be helpful. The similarity between them lies primarily in the nature of the abstract elements and not the elements as substances that can be isolated and observed.

On moving all the way to the right, a remarkable group of elements, the noble gases, is encountered, all of which were first isolated just before or at the turn of the 20th century. Their main property, rather paradoxically, at least when they were first isolated, was that they lacked chemical properties. These elements, consisting of helium, neon, argon, and krypton, were not even included in early periodic tables, since they were unknown and completely unanticipated. When they were discovered, their existence posed a formidable challenge to the periodic system, but one which was eventually successfully accommodated by the extension of the table to include a new group, now labelled as group 18.

Another block of elements, found at the foot of the modern table, consists of the rare earths that are commonly depicted as being literally disconnected. But this is just an apparent feature of this generally used display of the periodic system. Just as the transition metals are generally inserted as a block into the main body of the table, it is quite possible to do the same with the rare earths. Indeed, many such long-form displays have been published. While the long-form tables (Figure 6) give the rare earths a more natural place among the rest of the elements, they are rather cumbersome and do not readily lend themselves to conveniently shaped wall-charts of the periodic system. Although there are a number of different forms of the periodic table, what underlies the entire edifice, no matter the form of its representation, is the periodic law.

6. Long-form periodic table

The periodic law

The periodic law states that after certain regular but varying intervals, the chemical elements show an approximate repetition in their properties. For example, fluorine, chlorine, and bromine, which all fall into group 17, share the property of forming white crystalline salts of formula NaX with the metal sodium (where X is any halogen atom). This periodic repetition of properties is the essential fact that underlies all aspects of the periodic system.

This talk of the periodic law raises some interesting philosophical issues. First of all, periodicity among the elements is neither constant nor exact. In the generally used medium-long form of the periodic table, the first row has two elements, the second and third each contain eight, the fourth and fifth contain 18, and so on. This implies a varying periodicity consisting of 2, 8, 8, 18, etc., quite unlike the kind of periodicity one finds in the days of the week or notes in a musical scale. In these latter cases, the period length is constant, such as seven for the days of the week as well as the number of notes in a Western musical scale.

Among the elements, however, not only does the period length vary, but the periodicity is not exact. The elements within any column of the periodic table are not exact recurrences of each other. In this respect, their periodicity is not unlike the musical scale, in which one returns to a note denoted by the same letter, which sounds like the original note but is definitely not identical to it in that it is an octave higher.

The varying length of the periods of elements and the approximate nature of the repetition has caused some chemists to abandon the term 'law' in connection with chemical periodicity. Chemical periodicity may not seem as law-like as most laws of physics. However, it can be argued that chemical periodicity offers an example of a typically chemical law, approximate and complex, but still fundamentally displaying law-like behaviour.

Perhaps this is a good place to discuss some other points of terminology. How is a periodic table different from a periodic system? The term 'periodic system' is the more general of the two. The periodic system is the more abstract notion that holds that there is a fundamental relationship among the elements. Once it becomes a matter of displaying the periodic system, one can choose a three-dimensional arrangement, a circular shape, or any number of different two-dimensional tables. Of course, the term 'table' strictly implies a two-dimensional representation. So although the term 'periodic table' is by far the best known of the three terms, law, system, and table, it is actually the most restricted.

Reacting elements and ordering the elements

Much of what is known about the elements has been learned from the way they react with other elements and from their bonding properties. The metals on the left-hand side of the conventional periodic table are the complementary opposites of the non-metals, which tend to lie towards the right-hand side. This is so because, in modern terms, metals form positive ions by the loss of electrons, while non-metals gain electrons to form negative ions. Such oppositely charged ions combine together to form neutrally charged salts like sodium chloride or calcium bromide. There are further complementary aspects of metals and non-metals. Metal oxides or hydroxides dissolve in water to form bases, while non-metal oxides or hydroxides dissolve in water to form acids. An acid and a base react together in a 'neutralization' reaction to form a salt and water. Bases and acids, just like metals and non-metals from which they are formed, are also opposite but complementary.

Acids and bases have a connection with the origins of the periodic system since they featured prominently in the concept of equivalent weights, which was first used to order the elements. The equivalent weight of any particular metal, for example, was originally obtained from the amount of metal that reacts with a certain amount of a chosen standard acid. The term 'equivalent weight' was subsequently generalized to denote the amount of an element that reacts with a standard amount of oxygen. Historically, the ordering of the elements across periods was determined by equivalent weight, then later by atomic weight, and eventually by atomic number (explained below).

Chemists first began to make quantitative comparisons among the amounts of acids and bases that reacted together. This procedure was then extended to reactions between acids and metals. This allowed chemists to order the metals on a numerical scale according to their equivalent weight, which, as mentioned, is just the amount of the metal that combines with a fixed amount of acid.

Atomic weights, as distinct from equivalent weights, were first obtained in the early 1800s by John Dalton, who indirectly inferred them from measurements on the masses of elements combining together. But there were complications in this apparently simple method that forced Dalton to make assumptions about the chemical formulas of the compounds in question. The key to this question is the valence, or combining power, of an element. For example, a univalent atom combines with hydrogen atoms in a ratio of 1:1. Divalent atoms, like oxygen, combine in a ratio of 2:1, and so on.

Equivalent weight, as mentioned above, is sometimes regarded as a purely empirical concept since it does not seem to depend upon whether one believes in the existence of atoms. Following the introduction of atomic weights, many chemists who felt uneasy about the notion of atoms attempted to revert to the older concept of equivalent weights. They believed that equivalent weights would be purely empirical and therefore more reliable. But such hopes were an illusion, since equivalent weights also rested on the assumption of particular formulas for compounds, and formulas are theoretical notions.

For many years, there was a great deal of confusion created by the alternative use of equivalent weight and atomic weight. Dalton himself assumed that water consisted of one atom of hydrogen combined with one atom of oxygen, which would make its atomic weight and equivalent weight the same, but his guess at the valence of oxygen turned out to be incorrect. Many authors used the terms 'equivalent weight' and 'atomic weight' interchangeably, thus further adding to the confusion. The true relationship between equivalent weight, atomic weight, and valency was only clearly established in 1860 at the first major scientific conference, which was held in Karlsruhe, Germany. This clarification and the general adoption of consistent atomic weights cleared the path for the independent discovery of the periodic system by as many as six individuals in various countries, who each proposed forms of the periodic table that were successful to varying degrees. Each placed the elements generally in order of increasing atomic weight.

The third, and most modern, of the ordering concepts mentioned earlier is atomic number. Once atomic number was understood, it displaced atomic weight as the ordering principle for the elements. No longer dependent on combining weights in any way, atomic number can be given a simple microscopic interpretation in terms of the structure of the atoms of any element. The atomic number of an element is given by the number of protons, or units of positive charge, in the nucleus of any of its atoms. Thus each element on the periodic table has one more proton than the element preceding it. Since the number of neutrons in the nucleus also tends to increase as one moves through the periodic table, this makes atomic number and atomic weights roughly correspondent, but it is the atomic number that identifies any particular element. In other words, atoms of any particular element always have the same number of protons but they can differ in the number of neutrons they contain, a feature that produces the phenomenon of isotopy. These variants are called *isotopes*.

Different representations of the periodic system

The modern periodic system succeeds remarkably well in ordering the elements by atomic number in such a way that they fall into natural groups, but this system can be represented in more than one way. Thus there are many forms of the periodic table, some designed for different uses. Whereas a chemist might favour a form that highlights the reactivity of the elements, an electrical engineer might wish to focus on similarities and patterns in electrical conductivities.

The way in which the periodic system is displayed is a fascinating issue, and one that especially appeals to the popular imagination. Since the time of the early periodic tables of Newlands, Lothar Meyer, and Mendeleev, there have been many attempts to obtain the 'ultimate' periodic table. Indeed, it has been estimated that within one hundred years of the introduction of Mendeleev's famous table of 1869, approximately 700 different versions of the periodic table had been published. They included all kinds of alternatives, such as three-dimensional tables, helices, concentric circles, spirals, zigzags, step tables, mirror-image tables, and so on. Even today, articles are regularly published purporting to show new and improved versions of the periodic system.

What is fundamental to all these attempts is the periodic *law* itself, which exists in only one form. None of the multitude of displays changes this aspect of the periodic system. Many chemists stress that it does not matter how this law is physically represented, provided that certain basic requirements are met. Nevertheless, from a philosophical point of view, it is still relevant to consider the most fundamental representation of the elements, or the ultimate form of the periodic system, especially as this relates to the question of whether the periodic law should be regarded in a realistic manner or as a matter of convention. The usual response that representation is only a matter of convention would seem to clash with the realist notion that there may be a fact of the matter concerning the points at which the repetitions in properties occur in any periodic table.

Recent changes in the periodic table

In 1945, the American chemist Glen Seaborg suggested that the elements beginning with actinium, number 89, should be considered as a rare earth series, whereas it had previously been supposed that the new series of rare earths would begin after element number 92, or uranium (Figure 7). Seaborg's new periodic table revealed an analogy between europium (63) and gadolinium (64) and the as yet undiscovered elements 95 and 96 respectively. On the basis of these analogies, Seaborg succeeded in synthesizing and identifying the two new elements, which were subsequently named americium and curium. A number of further transuranium elements have subsequently been synthesized.

The standard form of the periodic table has also undergone some minor changes regarding the elements that mark the beginning of the third and fourth rows of the transition elements. Whereas older periodic tables show these elements to be lanthanum (57) and actinium (89), more recent experimental evidence and analysis have put lutetium (71) and lawrencium (103) in their former places (see Chapter 10). It is also interesting to note that some even older periodic tables based on macroscopic properties had anticipated these changes.

These are examples of ambiguities in what may be termed secondary classification, which is not as unequivocal as primary classification, or the sequential ordering of the elements. In classical chemical terms, secondary classification corresponds to the chemical similarities between the various elements in a group. In modern terms, secondary classification is explained using the concept of electronic configurations. Regardless of

First periodic table (before Seaborg's modification):

																	H	He
Li	Be											B	C	N	O	F	Ne
Na	Mg											Al	Si	P	S	Cl	Ar
K	Ca	Sc	Ti	V	Cr	Mn	Fe	Co	Ni	Cu	Zn	Ga	Ge	As	Se	Br	Kr
Rb	Sr	Y	Zr	Nb	Mo	Tc	Ru	Rh	Pd	Ag	Cd	In	Sn	Sb	Te	I	Xe
Cs	Ba	RE	Hf	Ta	W	Re	Os	Ir	Pt	Au	Hg	Tl	Pb	Bi	Po	At	Rn
Fr	Ra	Ac	Th	Pa	U

rare earths: | La | Ce | Pr | Nd | Pm | Sm | Eu | Gd | Tb | Dy | Ho | Er | Tm | Yb | Lu |

Second periodic table (after Seaborg's modification):

																	H	He
Li	Be											B	C	N	O	F	Ne
Na	Mg											Al	Si	P	S	Cl	Ar
K	Ca	Sc	Ti	V	Cr	Mn	Fe	Co	Ni	Cu	Zn	Ga	Ge	As	Se	Br	Kr
Rb	Sr	Y	Zr	Nb	Mo	Tc	Ru	Rh	Pd	Ag	Cd	In	Sn	Sb	Te	I	Xe
Cs	Ba	LA	Hf	Ta	W	Re	Os	Ir	Pt	Au	Hg	Tl	Pb	Bi	Po	At	Rn
Fr	Ra	AC

LA: | La | Ce | Pr | Nd | Pm | Sm | Eu | Gd | Tb | Dy | Ho | Er | Tm | Yb | Lu |

AC: | Ac | Th | Pa | U | Np | Pu |

7. **Periodic tables before and after Seaborg's modification**

whether one takes a classical chemical approach or a more physical approach based on electronic configurations, secondary classification of this type is more tenuous than primary classification and cannot be established as categorically. The way in which secondary classification, as defined here, is established is a modern example of the tension between using chemical properties or physical properties for classification. The precise placement of an element within groups of the periodic table can vary depending on whether one puts more emphasis on electronic configuration (a physical property) or its chemical properties. In fact, many recent debates on the placement of helium in the periodic system revolve around the relative importance that should be assigned to these two approaches (see Chapter 10).

In recent years, the number of elements has increased well beyond 100 as the result of the synthesis of artificial elements. At the time of writing, evidence has been reported for element number 117 and 118. Such elements are typically very unstable and only a few atoms are produced at any time. However, ingenious chemical techniques have been devised that permit the chemical properties of these so-called 'superheavy' elements to be examined and allow one to check whether extrapolations of chemical properties are maintained for such highly massive atoms. On a more philosophical note, the production of these elements allows us to examine whether the periodic law is an exceptionless law, of the same kind as Newton's law of gravitation, or whether deviations from the expected recurrences in chemical properties might take place once a sufficiently high atomic number is reached. No surprises have been found so far, but the question of whether some of these superheavy elements have the expected chemical properties is far from being fully resolved. One important complication that arises in this region of the periodic table is the increasing significance of relativistic effects (see below). These effects cause the adoption of unexpected electronic configurations in some atoms and may result in equally unexpected chemical properties.

Understanding the periodic system

Developments in physics have had a profound influence on the manner in which the periodic system is now understood. The two important theories in modern physics are Einstein's theory of relativity and quantum mechanics.

The first of these has had a limited impact on our understanding of the periodic system but is becoming increasingly important in accurate calculations carried out on atoms and molecules. The need to take account of relativity arises whenever objects move at speeds close to that of light. Inner electrons, especially those in the heavier atoms in the periodic system, can readily attain such relativistic velocities. It would be impossible to carry out an accurate calculation, especially on a heavy atom, without applying the necessary relativistic corrections. In addition, many seemingly mundane properties of elements, like the characteristic colour of gold or the liquidity of mercury, can best be explained as relativistic effects due to fast-moving inner-shell electrons.

But it is the second theory of modern physics that has exerted by far the more important role in attempts to understand the periodic system theoretically. Quantum theory was actually born in the year 1900. It was first applied to atoms by Niels Bohr, who pursued the notion that the similarities between the elements in any group of the periodic table could be explained by their having equal numbers of outer-shell electrons. The very notion of a particular number of electrons in an electron shell is an essentially quantum-like concept. Electrons are assumed to possess only certain quanta, or packets, of energy and, depending on how many such quanta they possess, they lie in one or another shell around the nucleus of the atom (see Chapter 7).

Soon after Bohr had introduced the quantum to the atom, many others developed his theory until the old quantum theory gave rise to quantum mechanics (see Chapter 8). Under the new description, electrons are regarded as much as waves as they are as particles. Even stranger is the notion that electrons no longer follow definite trajectories, or orbits, around the nucleus. Instead, the description changes to talk of smeared-out electron clouds, which occupy so-called orbitals. The most recent explanation of the periodic system is given in terms of how many such orbitals are populated by electrons. The explanation depends on the electron arrangement, or 'configuration', of an atom, which is spelled out in terms of the occupation of its orbitals.

The interesting question raised here is the relationship between chemistry and modern atomic physics, and in particular quantum mechanics. The popular view reinforced in most textbooks is that chemistry is nothing but physics 'deep down' and that all chemical phenomena, and especially the periodic system, can be developed on the basis of quantum mechanics. There are some problems with this view, however, which will be considered. For example, it will be suggested that the quantum mechanical explanation for the periodic system is still far from perfect. This is important because chemistry books, especially textbooks aimed at teaching, tend to give the impression that our current explanation of the periodic system is essentially complete. This is just not the case, as will be discussed.

The periodic table ranks as one of the most fruitful and unifying ideas in the whole of modern science, comparable perhaps with Darwin's theory of evolution by natural selection. After evolving for nearly 150 years through the work of numerous individuals, the periodic table remains at the heart of the study of chemistry. This is mainly because it is of immense practical benefit for making predictions about all manner of chemical and physical properties of the elements and possibilities for bond formation. Instead of having to learn the properties of the 100 or more elements, the modern chemist, or the student of chemistry, can make effective predictions from knowing the properties of typical members of each of the eight main groups and those of the transition metals and rare earth elements.

Having laid some thematic foundations and defined some key terms, we will now begin the story of the development of the modern periodic system, starting with its birth in the 18th and 19th centuries.

Chapter 3
Atomic weight, triads, and Prout

The first classifications of elements into groups were carried out on the basis of chemical similarities between elements, that is to say, on the basis of qualitative rather than quantitative aspects of the elements. For example, it was clear that the metals lithium, sodium, and potassium shared many similarities including their being soft, their floating on water, and the fact that unlike most metals, they could react visibly with water.

But the modern periodic table is based as much on quantitative properties of the elements as it is on their qualitative ones. Chemistry as a whole began to be a quantitative field, meaning *how much* reacts rather than *how* it reacts, as early as the 16th and 17th centuries. One of those responsible for this approach was Antoine Lavoisier, a French nobleman who was later guillotined in the aftermath of the French Revolution. Lavoisier was among the first to make accurate measurements of the weights of chemical reactants and their products. By doing this, Lavoisier was able to refute a long-standing assumption that a substance called 'phlogiston' was evolved when substances were burned.

On the contrary, Lavoisier discovered that burning any substance, such as an element, resulted in an increase rather than a decrease in weight. He also discovered that in any chemical operation, an equal quantity of matter exists before and after the operation. The discovery of this law of conservation of matter was followed by other laws of chemical combination, all of which began to demand a deeper explanation and one that would eventually lead to the discovery of the periodic table.

Lavoisier also turned away from the Greek notion of an abstract element as a bearer of properties. Instead, he concentrated on elements as the final stage in the decomposition of any compound. Although the notion of abstract elements would later return, in a modified form, it was necessary to make a clean break with the ancient Greek tradition, especially as many mysterious and non-scientific notions had continued to flourish in the Middle Ages and among the alchemists.

Returning to quantitative aspects of the elements, Benjamin Richter, working in Germany in 1792, published a list of what became known as equivalent weights (Figure 8). This was a list of the weights of various metals that reacted with a fixed amount of a particular acid, nitric acid for example. Now for the first time, the properties of various elements could be compared with each other in a simple quantitative way.

Dalton

In 1801, a young schoolteacher in Manchester, England, published the beginnings of a modern atomic theory. Following in the new tradition of Lavoisier and Richter, John Dalton took the ancient Greek notion of atoms, or the smallest particles of any substance, and made it quantitative. Not only did he assume that each element consisted of a particular type of atom, but he also began to estimate their relative weights.

For example, he drew on Lavoisier's experiments on the combination of hydrogen and oxygen to form water. Lavoisier had

Base		Acid	
Alumina	525	Hydrofluoric	427
Magnesia	615	Carbonic	577
Ammonia	672	Sebaic	706
Lime	793	Muriatic	712
Soda	859	Oxalic	755
Strontia	1329	Phoshproic	979
Potash	1605	Sulphuric	1000
Baryta	2222	Succinic	1209
		Nitric	1405
		Acetic	1480
		Citric	1583
		Tartaric	1694

8. **Richter's table of equivalent weights, as modified by Fischer in 1802**

shown that water was composed of 85% oxygen and 15% hydrogen. Dalton proposed that water consisted of one atom of hydrogen combined with one of oxygen to give a formula of HO, and that the atomic weight of oxygen was therefore $85/15 = 5.66$, assuming the weight of a hydrogen atom to be one unit. The current value for the oxygen atom is actually 16, and the difference lies in two issues that Dalton was unaware of. First of all, he was incorrect to assume that water is HO since, as everybody now knows, the formula is rather H_2O. Secondly, Lavoisier's data were not very accurate.

Dalton's notion of atomic weights gave a very reasonable explanation for the law of constant proportion, namely the fact that when two elements combine together, they do so in a constant ratio of their weights. This law could now be regarded as a scaled-up version of the combination of two or more atoms having particular atomic weights. The fact that macroscopic samples consist of a fixed ratio by weight of two elements reflects the fact that two particular atoms are combining many times over and, since they have particular masses, the product will also reflect that mass ratio.

Other chemists, as well as Dalton himself, discovered yet another law of chemical combination, the law of multiple proportions. When one element A combines with another one, B, to form more than one compound, there is a simple ratio between the combining masses of B in the two compounds. For example, carbon and oxygen combine together to form carbon monoxide and carbon dioxide. The weight of combined oxygen in the dioxide is twice as much as the weight of combined oxygen in the monoxide. Again, this law found a good explanation in Dalton's atomic theory because it implied that one atom of carbon had combined with one atom of oxygen in carbon monoxide CO, and two atoms of oxygen had combined with one of carbon in CO_2.

Von Humboldt and Gay-Lussac

Now consider one further law of chemical combination, but one which was not at first explained by Dalton's theory. In 1809, Alexander von Humboldt and Joseph-Louis Gay-Lussac discovered that when water vapour was formed by reacting two gases, hydrogen and oxygen, the volume of hydrogen was almost twice that of oxygen. Moreover, the volume of water vapour formed was about the same as that of the combining hydrogen.

2 volumes of hydrogen + 1 volume of oxygen→ 2 volumes of water vapour

This kind of behaviour was also found to apply to other gases combining together, so that von Humboldt and Gay-Lussac could conclude:

> The volumes of gases entering into chemical reaction and the gaseous products are in a ratio of small integers.

This new law of chemistry presented a major challenge to Dalton's new atomic theory. According to Dalton, any atom was absolutely indivisible, but this law could not be interpreted on the assumption of indivisible atoms of the gases concerned. It is only if the oxygen atoms are divisible that the above reaction between hydrogen and oxygen can possibly take place between two atoms of hydrogen and one of oxygen.

The solution to this puzzle came when an Italian physicist, Amedeo Avogadro, realized that it was rather two diatomic molecules of hydrogen that were combining with one diatomic molecule of oxygen. Nobody had previously thought that these gases consisted of two atoms of the element combined together to form a diatomic molecule. Since these molecules consisted of two atoms, it was the molecules that were divisible not the atoms themselves. Dalton's theory and the indivisibility of atoms could still be upheld and, by assuming the existence of diatomic gas molecules, consisting of two atoms of the same element, the new law of von Humboldt and Gay-Lussac could also be explained.

The reaction between hydrogen and oxygen was such that two hydrogen diatomic molecules broke up to form four atoms while one oxygen diatomic molecule broke up to form two oxygen atoms. Two molecules of water vapour, or H_2O, would then form to account for all the six atoms in question. This all looks very simple in hindsight, but given that diatomic molecules represented a radical idea and that the formula for the water molecule was not known, it is not surprising that an equation as simple as,

$$2H_2 + O_2 \rightarrow 2H_2O$$

took about 50 years before it was fully understood.

But by an odd historical twist, Dalton himself refused to accept the idea of diatomic molecules because he firmly believed that any two atoms of the same element should repel each other and that as a result they could never form a diatomic molecule. The idea of a chemical bond between any two like atoms was new and took some getting used to, especially to somebody like Dalton who had a rather elaborate view of how atoms should behave. Meanwhile, somebody like Avogadro could forge ahead and postulate diatomic molecules while being unencumbered by the idea that two like atoms repel each other, which in fact they do not, as we realize today.

Avogadro's idea of the formation of diatomic molecules was also arrived at independently by André Ampère, after whom the amp, or unit of current, is now named. But this crucial discovery lay dormant for something like 50 years until it was finally resuscitated by another Italian, Stanislao Cannizzaro, who lived in Sicily.

Prout's hypothesis

A few years after Dalton and others had begun publishing lists of atomic weights, the Scottish physician William Prout noticed something rather interesting. Many of the atomic weights that had been determined for the elements seemed to be whole-number multiples of the weight of the hydrogen atom. His conclusion was rather an obvious one. Perhaps all the atoms were simply composed of atoms of hydrogen. If this was true, it would also suggest the unity of all matter at the fundamental level, a notion that has been toyed with since the dawn of Greek philosophy and has resurfaced many times in different forms.

But not all the published atomic weights were exact multiples of the weight of hydrogen. Prout was not put off by this feature but suggested that the cause lay in the fact that the weights of these aberrant atoms had not yet been accurately determined. Prout's hypothesis, as it came to be known, was therefore rather productive because it caused others to measure atomic weights more and more accurately in order to prove him either right or wrong. The ensuing more accurate atomic weights would eventually play a pivotal role in the discovery and evolution of the periodic table.

But the initial consensus regarding Prout's hypothesis was that it was incorrect. The more accurately measured atomic weights suggested that, in general, atoms were not multiples of hydrogen atoms. Nevertheless, Prout's hypothesis was destined to make a comeback a good deal later in the story, although in a somewhat modified form.

Döbereiner's triads

The German chemist Wolfgang Döbereiner discovered another general principle that also contributed to the need to measure atomic weights more accurately and so paved the way for the periodic table. Starting in 1817, Döbereiner discovered the existence of various groups of elements in which one of the elements had both chemical properties and an atomic weight which was roughly the average of two other elements. These groups of three elements became known as triads. For example, lithium, sodium, and potassium are all softish grey metals with low densities. Lithium shows little reaction with water, whereas potassium is highly reactive. But sodium shows intermediate reactivity between that of these two other members of the triad.

In addition, the atomic weight of sodium (23) is intermediate between those of lithium (7) and potassium (39). This discovery was highly significant because it gave the first hint of a numerical regularity lying at the heart of the relationship between the nature and properties of the elements. It suggested an underlying mathematical order as to how the elements were related to each other chemically.

Another key triad recognized by Döbereiner consisted of three halogen elements, chlorine, bromine, and iodine. But Döbereiner did not attempt to connect these different triads together in any way. Had he done so, he might have discovered the periodic table some 50 years before Mendeleev and others did.

In identifying several triads, Döbereiner demanded that there should be a chemical similarity among the three elements in question in addition to the mathematical relationship mentioned. Other authors who followed him were not as fussy as him on the first point, and some of them believed that they had discovered many other triads. For example, in 1857 a 20-year-old German chemist, Ernst Lenssen, working in Wiesbaden, published an article in which he arranged virtually all of the 58 known elements into a total of 20 triads. Ten of his triads consisted of non-metals and acid-forming metals and the remaining ten of just metals.

Using the 20 triads in the table below, Lenssen also claimed to identify a total of 7 super-triads, in which the mean equivalent weight of each middle triad lies approximately midway between the mean weights of the other triads in a group of three triads. These were triads of triads, as it were. But Lenssen's system was somewhat forced. For example, in place of a genuine triad, he counted just one element, hydrogen, as constituting a triad because he found it convenient to do so. Furthermore, many of the triads that he claimed to exist looked appealing in numerical terms, but had no chemical significance. Lenssen and some other chemists became seduced by apparent numerical regularities while leaving chemistry behind.

	Calculated atomic weight			Determined atomic weights		
1	(K + Li)/2	= Na	= 23.03	39.11	23.00	6.95
2	(Ba + Ca)/2	= Sr	= 44.29	68.59	47.63	20
3	(Mg + Cd)/2	= Zn	= 33.8	12	32.5	55.7
4	(Mn + Co)/2	= Fe	= 28.5	27.5	28	29.5
5	(La + Di)/2	= Ce	= 48.3	47.3	47	49.6
6	Yt Er Tb			32	?	?
7	Th norium Al			59.5	?	13.7
8	(Be + Ur)/2	= Zr	= 33.5	7	33.6	60
9	(Cr + Cu)/2	= Ni	= 29.3	26.8	29.6	31.7
10	(Ag + Hg)/2	= Pb	= 104	108	103.6	100
11	(O + C)/2	= N	= 7	8	7	6
12	(Si + Fl)/2	= Bo	= 12.2	15	11	9.5
13	(Cl + J)/2	= Br	= 40.6	17.7	40	63.5
14	(S + Te)/2	= Se	= 40.1	16	39.7	64.2
15	(P + Sb)/2	= As	= 38	16	37.5	60
16	(Ta + Ti)/2	= Sn	= 58.7	92.3	59	25
17	(W + Mo)/2	= V	= 69	92	68.5	46
18	(Pa + Rh)2	= Ru	= 52.5	53.2	52.1	51.2
19	(Os + Ir)/2	= Pt	= 98.9	99.4	99	98.5
20	(Bi + Au)/2	= Hg	= 101.2	104	100	98.4

9. The 20 triads of Lenssen

Another system of classification of the elements was due to
Leopold Gmelin working in Germany in 1843. This author
discovered some new triads and did begin to connect them
together to form an overall classification system with a rather
distinctive shape (Figure 10). His system contained as many as 55
elements, and in anticipation of later systems, it appears that
Gmelin ordered most of the elements in order of increasing
atomic weights, although never explicitly expressing this notion.

However, Gmelin's system cannot be regarded as a periodic system
since it does not display the repetition in the properties of the
elements. In other words, the property of chemical periodicity
from which the periodic table derives its name was not yet
featured. Gmelin went on to use his system of elements in order to
organize a textbook of chemistry consisting of 500 or so pages.

```
         O                    N                      H
 F  Cl  Br  I                              Li  Na  K
    S  Se  Te                          Mg  Ca  Sr  Ba
    P  As  Sb                          Be  Ce  La
    C  B  Bi                           Zr  Th  Al
    Ti  Ta  W              Sn  Cd  Zn
       Mo  V  Cr        U  Mn  Ni  Fe
          Bi  Pb  Ag  Hg  Cu
          Os  Ir  Rh  Pt  Pd  Au
```

10. Gmelin's table of elements

This is probably the first time that a table of elements was used as a basis for an entire book on chemistry, something that is completely standard these days, although we should recall that this was not a *periodic* table.

Kremers

A modern periodic table is much more than a collection of groups of elements showing similar chemical properties. In addition to what may be called 'vertical relationships', which embody triads of elements, a modern periodic table connects together groups of elements into an orderly sequence.

A periodic table consists of a horizontal dimension, containing dissimilar elements, as well as a vertical dimension with similar elements. The first person to consider a horizontal relationship was Peter Kremers from Cologne in Germany. He noted the following regularity among a short series of elements that included oxygen, sulphur, titanium, phosphorus, and selenium (Figure 11).

	O	S	Ti	P	Se
Atomic weight	8	16	24.12	32	39.62
difference		8	8	~8	~8

11. Kremers' atomic weight differences for the oxygen series

Kremers also discovered some new triads such as

$$Mg = \frac{O+S}{2}, \quad Ca = \frac{S+Ti}{2}, \quad Fe = \frac{Ti+P}{2}$$

From a modern standpoint, these triads may not seem to be chemically significant. But this is because the modern medium-long form of the periodic system fails to display secondary kinships among some elements. Sulphur and titanium both show a valency of four, for example, though they do not appear in the same group in the medium-long form of the periodic system. But it is not so far-fetched to consider them as being chemically analogous. Given the fact that both titanium and phosphorus commonly display valences of three, this grouping too is not as incorrect as a modern reader may think. But broadly speaking, as in the case of Lenssen, this is a desperate attempt to create new triads at all costs. The aim seems to have become one of finding triad relationships among the weights of the elements irrespective of whether or not these had any chemical significance. Mendeleev later described such activity among his colleagues as an obsession with triads, which he believed had delayed the discovery of the mature periodic system.

But to return to Kremers, his most incisive contribution lay in the suggestion of a bi-directional scheme of what he termed 'conjugated triads'. Here, certain elements would serve as members of two distinct triads lying perpendicularly to each other,

Li 6.5	Na 23	K 39.2
Mg 12	Zn 32.6	Cd 56
Ca 20	Sr 43.8	Ba 68.5

Thus, in a more profound way than any of his predecessors, Kremers was comparing chemically *dissimilar* elements, a practice which would only reach full maturity with the tables of Lothar Meyer and Mendeleev.

Chapter 4
Steps towards the periodic table

The 1860s was an important decade for the discovery of the periodic table. It began with a congress convened in Karlsruhe, Germany, aimed at resolving a number of technical issues having to do with chemists' understanding of the concept of atoms and molecules.

As mentioned in Chapter 2, the law of combining gas volumes discovered by Gay-Lussac could only be explained by assuming the existence of divisible diatomic molecules of two or more combined atoms, such as H_2, O_2, etc. This proposal had still not been generally accepted because of criticisms from Dalton, among others. At the Karlsruhe conference, the idea finally gathered widespread acceptance due to the advocacy of Cannizzaro, a fellow countryman of Avogadro who had first suggested the idea 50 years previously.

Another problem had been that many different authors gave different values of the atomic weights of the elements. Cannizzaro also succeeded in producing a rationalized set of values that he printed in a small pamphlet that was distributed to delegates departing from the conference. Armed with these reforms, it was just a matter of time before as many as six individual scientists developed rudimentary periodic systems which included most of the then known 60 or so elements.

De Chancourtois

The first person to actually discover chemical periodicity was a French geologist, Émile Béguyer De Chancourtois, who arranged the elements in order of increasing atomic weights on a spiral that he inscribed around a metal cylinder. After doing so, he noticed that chemically similar elements fell on vertical lines, which intersected the spiral as it circled the cylinder (Figure 12). Here was the essential discovery that the elements, when arranged in their natural order, seemed to recur approximately after certain regular intervals. Just like the days of the week, the months of the year, or the notes in a musical scale, periodicity or repetition seems to be an essential property of the elements. The underlying cause of chemical repetition would remain mysterious for many years to come.

De Chancourtois expressed his support for Prout's hypothesis and went as far as rounding off the values of atomic weights to whole numbers in his periodic system. Sodium, with a weight of 23, appeared in his system at one whole turn away from lithium, with an atomic weight of 7. In the next column, he placed the elements magnesium, calcium, iron, strontium, uranium, and barium. In the modern periodic system, four of these elements, magnesium, calcium, strontium, and barium, are indeed in the same group. The inclusion of iron and uranium seems at first to be an outright error, but as we will see, many early short-form periodic tables featured some transition elements among what are now termed the main group elements.

The Frenchman was also somewhat unlucky in that his first and important publication failed to include a diagram of his system, a rather fatal omission given the fact that graphical representation is *the* crucial aspect of any periodic system. In order to remedy this problem, De Chancourtois later republished his paper privately, but as a result it did not gain widespread distribution and

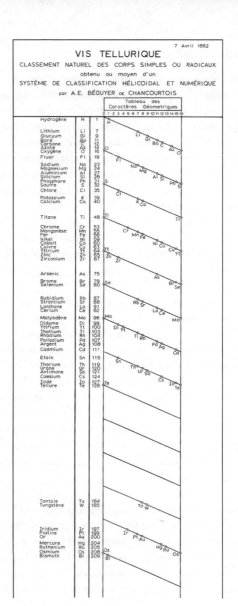

12. De Chancourtois' telluric screw

remained obscured from the chemical community of the day. Chancourtois' key discovery was also not noticed by the world's chemists because he was not one of them, but was a geologist. Finally, the discovery was not noticed because it was ahead of its time.

In fact, even after Mendeleev had achieved a good deal of fame for his own periodic system, which he began to publish in 1869, most chemists had not yet heard of De Chancourtois' work, in his native France or anywhere else. Then finally in 1892, 30 years after De Chancourtois' path-breaking paper had appeared, three chemists went to some lengths to try to resurrect his work.

In England, Philip Hartog became irritated after hearing Mendeleev saying that De Chancourtois had not regarded his own system as being a natural one. Hartog then published a paper in support of De Chancourtois. Meanwhile in France, Paul Émile Le Coq De Boisbaudran and Albert Auguste Lapparent made similar pleas on behalf of their countryman's priority and restored at least a little Gallic honour.

Newlands

John Newlands was a London-based sugar chemist whose mother was of Italian birth, a fact that seems to have induced him to volunteer to fight in Garibaldi's revolutionary campaign to unify Italy. In any case, no harm came to young Newlands since he was soon back at work in London. In 1863, just one year after De Chancourtois' paper, Newlands published his first classification of elements. Without using the atomic weights of Cannizzaro, of which he was unaware, Newlands placed the then known elements into eleven groups whose members showed analogous properties. In addition, he noted that their atomic weights differed by a factor of eight or some multiple of eight (Figure 13).

Group I. Metals of the alkalies:—Lithium, 7; sodium, 23; potassium, 39; rubidium, 85; cæsium, 123; thallium, 204.

The relation among the equivalents of this group (see CHEMICAL NEWS, January 10, 1863) may, perhaps, be most simply stated as follows:—

1 of lithium + 1 of potassium = 2 of sodium.

1	,,	+ 2	,,	= 1 of rubidium.
1	,,	+ 3	,,	= 1 of cæsium.
1	,,	+ 4	,,	= 163, the equivalent of a metal not yet discovered.
1	,,	+ 5	,,	= 1 of thallium.

Group II. Metals of the alkaline earths:—Magnesium, 12; calcium, 20; strontium, 43·8; barium, 68·5.

In this group, strontium is the mean of calcium and barium.

Group III. Metals of the earths:—Beryllium, 6·9; aluminium, 13·7; zirconium, 33·6; cerium, 47; lanthanium, 47; didymium, 48; thorium, 59·6.

Aluminium equals two of beryllium, or one-third of the sum of beryllium and zirconium. (Aluminium also is one-half of manganese, which, with iron and chromium, forms sesquioxides, isomorphous, with alumina.)

1 of zirconium + 1 of aluminium = 1 of cerium.

| 1 | ,, | + 2 | ,, | = 1 of thorium. |

Lanthanium and didymium are identical with cerium, or nearly so.

Group IV. Metals whose protoxides are isomorphous with magnesia:—Magnesium, 12; chromium, 26·7; manganese, 27·6; iron, 28; cobalt, 29·5; nickel, 29·5; copper, 31·7; zinc, 32·6; cadmium, 56.

Between magnesium and cadmium, the extremities of this group, zinc is the mean. Cobalt and nickel are identical. Between cobalt and zinc, copper is the mean. Iron is one-half of cadmium. Between iron and chromium, manganese is the mean.

Group V.—Fluorine, 19; chlorine, 35·5; bromine, 80; iodine, 127.

In this group bromine is the mean between chlorine and iodine.

Group VI.—Oxygen, 8; sulphur, 16; selenium, 39·5; tellurium, 64·2.

In this group selenium is the mean between sulphur and tellurium.

Group VII.—Nitrogen, 14; phosphorus, 31; arsenic, 75; osmium, 99·6; antimony, 120·3; bismuth, 213.

13. First seven of Newlands' grouping of elements in 1863

For example, his group I consisted of lithium (7), sodium (23), potassium (39), rubidium (85), caesium (123), and thallium (204). From a modern perspective, he had only placed thallium incorrectly since it belongs instead with boron, aluminium, gallium, and indium. The element thallium had been discovered just one year previously by the Englishman William Crookes. The first person to place it correctly in the boron group was a co-discoverer of the periodic system, namely Julius Lothar Meyer working in Germany. Even the great Mendeleev misplaced thallium in his early periodic tables and, like Newlands, he placed the element among the alkali metals.

In his first paper on classification of the elements, Newlands makes the following remark about the alkali metal group:

> The relation among the equivalents of this group may perhaps most simply be stated as follows, 1 of lithium (7) + 1 of potassium (39) = 2 of sodium.

This is, of course, nothing but a rediscovery of the triad relationship among these elements since,

Li   7

Na 23 $2\,Na\,(23) = 7 + 39$

K   39

In 1864, Newlands began publishing a series of articles in which he groped his way to a better periodic system and to what he later called the law of octaves, that is, the idea that elements repeat after moving through a sequence of eight. In 1865, he included 65 elements in his system and used ordinal numbers rather than atomic weights as a means of ordering the elements in a sequence of increasing weights. Now he began to write quite confidently of a new law, whereas De Chancourtois had only briefly considered this possibility but had rejected it.

No.	No.	No.	No.	No.	No.	No.	No.
H 1	F 8	Cl 15	Co & Ni 22	Br & Ni 22	Pd 36	I 42	Pt & Ir 50
Li 2	Na 9	K 16	Cu 23	Rb 30	Ag 37	Cs 44	Os 51
G 3	Mg 10	Ca 17	Zn 24	Sr 31	Cd 38	Ba & V 45	Hg 52
Bo 4	Al 11	Cr 19	Y 25	Ce & La 33	U 40	Ta 46	Tl 53
C 5	Si 12	Ti 18	In 26	Zr 32	Sn 39	W 47	Pb 54
N 6	P 23	Mn 20	As 27	Di & Mo 34	Sb 41	Nb 48	Bi 55
O 7	S 14	Fe 21	Se 28	Ro & Ru 35	Te 43	Au 49	Th 56

14. Newlands' table illustrating the law of octaves as presented to the Chemical Society in 1866

Partly because of this choice of a musical analogy in discussing 'octaves' and because Newlands was not an academic chemist, his idea was ridiculed when he presented it verbally to the Royal Society of Chemistry in 1866 (Figure 14). One member of this austere audience asked Newlands whether he had considered ordering the elements in alphabetical order. Newlands' paper was not published in the proceedings of the society, although he published further articles in a few other chemical journals. The time for the acceptance of the idea had still not come even though it was starting to occur to some chemists. But Newlands soldiered on and responded to his critics, while publishing a number of other periodic tables.

Odling

Another chemist to publish an early periodic table was William Odling, who, unlike Newlands, was a leading academic chemist. Odling attended the Karlsruhe conference and became a champion of Cannizzaro's views in Britain. He also held major appointments such as a professorship of chemistry at Oxford and the directorship of the Royal Institution in Albemarle Street in London. Odling independently published his own version of the periodic table by arranging the elements in order of increasing atomic weight, as Newlands had done, and by showing similar elements in vertical columns.

Cl	-	F	or	35.5	-	19	=	16.5
K	-	Na		39	-	23	=	16
Na	-	Li		23	-	7	=	16
Mo	-	Se		96	-	80	=	16
S	-	O		32	-	16	=	16
Ca	-	Mg		40	-	24	=	16
Mg	-	G		24	-	9	=	15
P	-	N		31	-	14	=	17
Al	-	B		27.5	-	11	=	16.5
Si	-	C		28	-	12	=	16

15. Odling's third table of differences

In a paper written in 1864, Odling states that:

> Upon arranging the atomic weights or proportional numbers of the
> sixty or so recognized elements in order of their several magnitudes,
> we observe a marked continuity in the resulting arithmetical series.

This is followed by the further statement:

> With what ease this purely arithmetical seriation may be made to
> accord with a horizontal arrangement of the elements according to
> their usually received groupings, is shown in the following table, in
> the first three columns of which the numerical sequence is perfect,
> while in the other two the irregularities are but few and trivial.

It is not clear why Odling's discovery was not accepted, since he
did not lack academic credentials. It seems to have been due to the
fact that Odling himself lacked enthusiasm for the idea of
chemical periodicity and was reluctant to believe that it might
represent a law of nature.

Hinrichs

In the United States, a newly arrived Danish immigrant, Gustavus
Hinrichs, was busy developing his own system of classification of

the elements, which he published in a striking radial format. However, Hinrichs' writing tended to be shrouded in layers of mysterious allusions to Greek mythology and other eccentric excesses, not to mention the fact that he managed to alienate himself from colleagues and the mainstream community of chemists.

Hinrichs was born in 1836 in Holstein, then a part of Denmark but later becoming a German province. He published his first book at the age of 20, while attending the University of Copenhagen. He emigrated to the United States in 1861 to escape political persecution, and after a year of teaching in a high school was appointed head of modern languages at the University of Iowa. A mere one year later, he became Professor of Natural Philosophy, Chemistry, and Modern Languages. He is also credited with founding the first meteorological station in the USA, and acted as its director for 14 years. One of the few detailed accounts of Hinrich's life and work was published by Karl Zapffe, who says:

> It is not necessary to read far into Hinrichs' numerous publications to recognize the marks of an egocentric zeal which defaced many of his contributions with an untrustworthy eccentricity. Only at this late date does it become possible to separate those inspirations which were real – and which swept him off his feet – from background material which he captured in the course of his own learning. Whatever the source, Hinrichs usually dressed it with multilingual ostentation, and to such a point of disguise that he even came to regard Greek philosophy as his own.

Hinrichs' wide range of interests extended to astronomy. Like many authors before him, as far back as Plato, Hinrichs noticed some numerical regularities regarding the sizes of the planetary orbits. In an article published in 1864, he showed the table in Figure 16, which he proceeded to interpret.

Hinrichs expressed the differences in these distances by the formula $2^x \times n$, in which n is the difference in the distances of

Distance to the Sun	
Mercury	60
Venus	80
Earth	120
Mars	200
Asteroid	360
Jupiter	680
Saturn	1320
Uranus	2600
Neptune	5160

16. **Hinrichs' table of planetary distances (1864)**

Venus and Mercury from the Sun, or 20 units. Depending on the value of x, the formula therefore gives the following distances:

$$2^0 \times 20 = 20$$
$$2^1 \times 20 = 40$$
$$2^2 \times 20 = 80$$
$$2^3 \times 20 = 160$$
$$2^4 \times 20 = 320$$
etc.

A few years previously, in 1859, the Germans Gustav Kirchoff and Robert Bunsen had discovered that each element could be made to emit light, which could then be dispersed with a glass prism and analysed quantitatively. What they also discovered was that every single element gave a unique spectrum consisting of a set of specific spectral lines, which they set about measuring and publishing in elaborate tables. Some authors suggested that these spectral lines might provide information about the various elements that had produced them, but these suggestions met with strenuous criticism from one of their discoverers, Bunsen. Indeed, Bunsen remained quite opposed to the idea of studying spectra in order to study atoms or to classify them in some way.

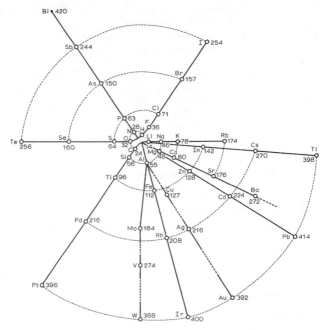

17. Hinrichs' periodic system

Hinrichs, however, had no hesitation in connecting spectra with the atoms of the elements. In particular, he became interested in the fact that, with any particular element, the frequencies of its spectral lines always seemed to be whole-number multiples of the smallest difference. For example, in the case of calcium, a ratio of 1:2:4 had been observed among its spectral frequencies. Hinrichs' interpretation of this fact was bold and elegant: if the sizes of planetary orbits produce a regular series of whole numbers, as mentioned earlier, and if the ratios among spectral line differences also produce whole-number ratios, the cause of the latter might lie in the size ratios among the atomic dimensions of the various elements.

after a priority dispute later developed with Mendeleev. Had this table been brought to light at that time, it is not clear that Mendeleev's claim to precedence would have been afforded the weight that it is given today.

Lothar Meyer's lost table has great merit for its inclusion of so many elements and for certain placements which Mendeleev's famous table of the same year failed to achieve. The lost table was eventually published, after Lothar Meyer's death, in 1895 and far too late to have any impact on the question of who had been the first to arrive at the first fully mature periodic system.

During the course of the rather public dispute, Mendeleev took a more forceful approach, claiming that the credit belonged to himself alone because he had gone beyond discovering the periodic system by making a number of successful predictions. Lothar Meyer appears to have adopted something of a defeatist attitude by even admitting that he had lacked the courage to make predictions.

Chapter 5
The Russian genius – Mendeleev

Dimitri Ivanovich Mendeleev is by far the most famous Russian scientist of the modern era. Not only did he discover the periodic system, but he recognized that it pointed to a deep law of nature, the periodic law. He also spent many years drawing out the full consequences of this law, most notably by predicting the existence and properties of many new elements. In addition, he corrected the atomic weights of some already known elements and successfully changed the position of some other elements in the periodic table.

But perhaps most important of all, Mendeleev made the periodic table his own by pursuing its study and development during several periods of his life, even though he worked in a number of other very diverse fields. By contrast, most of his precursors or co-discoverers failed to follow up their initial discoveries. As a result, the name of Mendeleev is inextricably linked with the periodic table in much the same way that evolution by natural selection and relativity theory are linked with Darwin and Einstein respectively. Given Mendeleev's supreme importance in the story of the periodic table, this entire chapter will be devoted to his scientific work as well as his early development.

When Mendeleev was very young, his father, who owned a glass factory, went blind and died soon afterwards. Dimitri, the youngest of fourteen siblings, was brought up by his adoring mother, who was determined to provide the best possible education for him. This devotion included going on a trip over hundreds of miles with the young man in a failed attempt to enrol him into Moscow University. The reason for Mendeleev's being rejected was apparently his Siberian origin and the fact that the university only admitted Russians. Undeterred, Mendeleev's mother succeeded in getting the young man into the Main Pedagogical Institute of St Petersburg, where he began to study chemistry, physics, biology, and of course pedagogy, the last of which was to have a telling effect on his discovering the mature periodic system. Sadly, Mendeleev's mother died very soon after he had entered this institution, and he was left to fend for himself.

After completing his undergraduate education, Mendeleev spent some time in France, and later Germany, where he was formally attached to the laboratory of Robert Bunsen, although he preferred to stay at home and conduct his private experiments on gases. It was during this period in Germany that Mendeleev attended the Karlsruhe conference in 1860, not because he was a prominent chemist but more because he happened to be at the right place at the right time. Although Mendeleev quickly grasped the value of Cannizzaro's ideas, as Lothar Meyer did at this conference, Mendeleev's conversion to using Cannizzaro's atomic weights appears to have taken considerably more time than it did for Lothar Meyer.

In 1861, Mendeleev began to show real promise when he published a textbook on organic chemistry which won him the coveted Demidov Prize in Russia. In 1865, he defended his doctoral thesis on the interaction between alcohol and water, and began to work on a book on inorganic chemistry in order to improve the teaching of chemistry. In the first volume of this new book, he treated the more common elements in no particular order. By 1868, he had completed this volume and had begun to consider how he should make the transition to the remaining elements in a second volume.

The actual discovery

Although Mendeleev had been thinking about elements, atomic weights, and classification for about ten years, he appears to have had his eureka moment, or perhaps 'eureka day', on 17 February 1869. On this day, he cancelled his trip to visit a cheese factory as a consultant and decided to work on what was to become his famous brain-child – the periodic table.

First, he listed the symbols for a handful of elements in two rows on the back of the invitation to the cheese factory.

Na	K	Rb	Cs
Be	Mg	Zn	Cd

Then he produced a slightly larger array of 16 elements.

F	Cl	Br	I				
Na	K	Rb	Cs			Cu	Ag
Mg	Ca	Sr	Ba	Zn	Cd		

By that evening, Mendeleev had sketched out an entire periodic table which included 63 known elements. Moreover, it included several gaps for as yet unknown elements and even the predicted atomic weights for some of these elements. Two hundred copies of this first table were printed and sent to chemists all over Europe. On 6 March of the same year, the discovery was announced by one of Mendeleev's colleagues at a meeting of the Russian Chemical Society. Within a month, an article had appeared in the journal of this newly formed society and another longer article appeared in Germany.

Many popular books and documentary programmes about Mendeleev claim that he arrived at his periodic table in the course of a dream or while laying out cards for each of the elements as though playing the game of patience. The second of these tales especially is now regarded as being apocryphal by Mendeleev biographers such as science historian Michael Gordin.

To turn to Mendeleev's scientific approach, it seems that he differed considerably from his rival Lothar Meyer in not believing in the unity of all matter and similarly not supporting Prout's hypothesis for the composite nature of all elements. Mendeleev also took care to distance himself from the notion of triads of elements. For example, he proposed that the element fluorine should be grouped together with chlorine, bromine, and iodine even though this would imply going beyond a triad to form a group of at least four elements.

Whereas Lothar Meyer had concentrated on physical principles, and mainly on the physical properties of the elements, Mendeleev was very familiar with their chemical properties. On the other hand, when it came to deciding upon the most important criterion for classifying the elements, Mendeleev insisted that atomic weight ordering would tolerate no exceptions. Of course, many of Mendeleev's precursors, such as De Chancourtois, Newlands, Odling, and Lothar Meyer, had recognized the importance of atomic weight ordering to varying degrees. But Mendeleev also reached a deeper philosophical understanding of atomic weights and the nature of elements which allowed him to move into uncharted territory regarding as yet unknown elements.

The nature of elements

There is a long-standing puzzle in chemistry. When sodium and chlorine combine, they give rise to a completely new substance, sodium chloride, in which the combining elements do not seem to survive, at least from a visual perspective. This is the phenomenon of chemical bonding, or chemical combination, which differs markedly from the physical mixing of, say, powdered sulphur and iron filings.

One of the questions, in the case of chemical combination, is to understand how, if at all, the combining elements survive in the compound. This issue is further complicated in several languages, such as English, in which we use the word 'element' to refer to the combined substance such as chlorine when it is present in sodium chloride. Now what underlies both the uncombined green gas chlorine and combined chlorine is sometimes also called 'element'. We now have three senses of the same central chemical term to describe the substances that the periodic table is supposed to classify.

The third sense of 'element', as introduced above, has been variously called metaphysical element, abstract element, transcendental element, and more recently 'element as basic substance'. This is the element as an abstract bearer of properties but lacking such phenomenal properties as the green colour of chlorine. Green chlorine, meanwhile, is said to be the element as a 'simple substance'.

When Antoine Lavoisier revolutionized chemistry at the end of the 17th century, one of his contributions was to concentrate attention onto the element as a simple substance, that is, the element in isolated form. This was intended to improve chemistry by ridding it of excess metaphysical baggage and it was indeed a great step forward. Now an 'element' was to be regarded primarily as the final step in the separation of the components in any compound. Whether Lavoisier intended to banish the more abstract and philosophical sense of 'element' is a moot point, but certainly this sense began to take a back-seat to the sense of an element that could be isolated.

However, the more abstract sense was not completely forgotten, and Mendeleev was one of the chemists who not only understood it but proposed to elevate its status. In fact, he repeatedly claimed that the periodic system was primarily a classification of this more abstract sense of the term 'element' and not necessarily the more concrete element that could be isolated.

The reason why I have carefully developed this issue is that, when he was armed with this notion, Mendeleev could take a deeper view of elements than chemists who restricted themselves to elements in isolated form. It afforded Mendeleev the possibility of going beyond appearances. If a particular element did not appear to fit within a particular group, Mendeleev could draw upon the deeper sense of the term 'element', and thus to some extent he could ignore the apparent properties of the element as an isolated or simple substance.

Mendeleev's predictions

One of his greatest triumphs, and perhaps the one that he is best remembered for, is Mendeleev's correct prediction of the existence of several new elements. In addition, he corrected the atomic weights of some elements as well as relocating other elements to new positions within the periodic table. As I suggested in the previous section, such far-sightedness may have been due to his having a greater philosophical understanding of the nature of the elements than his competitors. By concentrating on the more abstract concept of elements as basic substances, Mendeleev was able to surmount apparent obstacles that arose from taking the properties of the isolated elements at face value.

Although Mendeleev placed the greatest importance on the atomic weights of the elements he also considered chemical and physical properties and family resemblances among them. Whereas Lothar Meyer concentrated on physical properties, Mendeleev paid more attention to the chemistry of the elements. Another criterion he used was that each element should occupy a single place in the periodic table, although he was willing to violate this notion when it came to what he called group VIII (Figure 5). His more important criterion however was the ordering of the elements according to increasing atomic weights. In one or two cases, he appeared to even violate this principle, although a closer inspection shows that this was not in fact the case.

The case of tellurium and iodine is one of only four pair reversals in the periodic system and the best known among them. These are elements which appear in the reverse order than they would according to increasing values of atomic weight (Te = 127.6, I = 126.9, and yet Te appears before I). Many historical accounts make a point of recounting how clever Mendeleev was to reverse the positions of these elements, thus putting chemical properties over and above the ordering according to atomic weight. Such a claim is incorrect in several respects. First of all, Mendeleev was by no means the first chemist to make this particular reversal. Odling, Newlands, and Lothar Meyer all published tables in which the positions of tellurium and iodine had been reversed, well before the appearance of Mendeleev's articles. Secondly, Mendeleev was not in fact placing a greater emphasis on chemical properties than on atomic weight ordering.

Mendeleev held to his criterion of ordering according to increasing atomic weight and repeatedly stated that this principle would tolerate no exceptions. His thinking regarding tellurium and iodine was rather that the atomic weights for one or both of these elements had been incorrectly determined, and that future work would reveal that, even on the basis of atomic weight ordering, tellurium should be placed before iodine. On this point, as in many instances that tend to go unreported, Mendeleev was wrong.

At the time when Mendeleev proposed his first periodic systems, the atomic weights for tellurium and iodine were thought to be 128 and 127 respectively. Mendeleev's belief that atomic weight was the fundamental ordering principle meant that he had no choice but to question the accuracy of these two values. This was because it was clear that in terms of chemical similarities, tellurium should be grouped with the elements in group VI and iodine with those in group VII, or in other words, that this pair of elements should be 'reversed'. Mendeleev continued to question the reliability of these atomic weights until the end of his life.

Initially, he doubted the atomic weight of tellurium while believing that that of iodine was essentially correct. Mendeleev began to list tellurium as having an atomic weight of 125 in some of his subsequent periodic tables. At one time, he asserted that the commonly reported value of 128 was the result of measurements having been made on a mixture of tellurium and a new element he called eka-tellurium. Motivated by these claims, the Czech chemist Bohuslav Brauner began a series of experiments in the early 1880s aimed at the re-determination of the atomic weight of tellurium. By 1883, he reported that the value for tellurium should be 125. Mendeleev was sent a telegram of congratulations by other participants present at the meeting at which Brauner had made this announcement. In 1889, Brauner obtained new results that seemed to further strengthen the earlier finding of Te = 125.

But in 1895, everything changed as Brauner himself began reporting a new value for tellurium that was greater than that of iodine, thus returning matters to their initial starting point. Mendeleev's response was now to begin to question the accuracy of the accepted atomic weight for iodine instead of tellurium. This time, he requested a re-determination for the atomic weight of iodine and hoped that its value would turn out to be higher. In some of his later periodic tables, Mendeleev even listed tellurium and iodine as both having atomic weights of 127. The problem would not be resolved until work by Henry Moseley in 1913 and 1914, who showed that the elements should be ordered according to atomic number rather than atomic weight. While tellurium has the higher atomic weight than iodine, it has a lower atomic number, and this is why it should be placed before iodine in full agreement with its chemical behaviour.

Although Mendeleev's predictions might seem quite miraculous, they were in fact based on careful interpolation between the properties of elements flanking the unknown elements. He began to make predictions in his very first publication on the periodic system in 1869, although he published a more detailed account of his forecasts in a long paper of 1871. He began by focusing on two gaps, one below aluminium and the other below silicon in his periodic table, which he provisionally called eka-aluminium and eka-boron, using the Sanskrit prefix meaning 'one' or 'one-like'. In his paper of 1869, he wrote:

> We must expect the discovery of yet unknown elements, e.g. elements analogous to Al and Si, with atomic weights 65–75.

In the autumn of 1870, he began to speculate about a third element, which would lie below boron in the table. He listed the atomic volumes of the three elements as

eka-boron	eka-aluminium	eka-silicon
15	11.5	13

In 1871, he predicted their atomic weights to be

eka-boron	eka-aluminium	eka-silicon
44	68	72

and he gave a detailed set of predictions on various chemical and physical properties of all three elements.

It took 15 years before the first of these predicted elements, later called gallium, was isolated. With a few very minor exceptions, Mendeleev's predictions were almost exactly correct. The accuracy of Mendeleev's predictions can also be clearly seen in the case of the element he called eka-silicon, later called germanium, after it had been isolated by the German chemist Clemens Winkler. It would seem that Mendeleev's only failure lies in the specific gravity of the tetrachloride of germanium (Figure 20).

Property	Eka-silicon 1871 prediction	Germanium discovered 1886
Relative atomic mass	72	72.32
Specific gravity	5.5	5.47
Specific heat	0.073	0.076
Atomic volume	13 cm^3	13.22 cm^3
Colour	dark grey	greyish white
Specific gravity of dioxide	4.7	4.703
Boiling point of tetrachloride	100°C	86°C
Specific gravity of tetrachloride	1.9	1.887
Boiling point of tetra ethyl deriv.	160°C	160°C

20. The predicted and observed properties of eka-silicon (germanium)

Mendeleev's failed predictions

But not all of Mendeleev's predictions were so dramatically successful, a feature that seems to be omitted from most popular accounts of the history of the periodic table. As Figure 21 shows, he was unsuccessful in as many as nine out of his eighteen published predictions, although perhaps not all of these predictions should be given the same weight. This is because some of the elements involved the rare earths which resemble each other very closely and which posed a major challenge to the periodic table for many years to come.

In addition, Mendeleev's failed predictions raise another philosophical issue. For a long time, historians and philosophers of science have debated whether successful predictions made in the course of scientific advances should count more, or less, than the successful accommodation of already known data. Of course, there is no disputing the fact that successful predictions carry a greater psychological impact since they almost imply that the scientist in question can foretell the future. But successful accommodation, or explanation of already known data, is no mean feat either, especially since there is usually more already known information to incorporate into a new scientific theory. This was

Element as given by Mendeleev.	Predicted A.W.	Measured A.W.	Eventual name
coronium	0.4	not found	not found
ether	0.17	not found	not found
eka-boron	44	44.6	scandium
eka-cerium	54	not found	not found
eka-aluminium	68	69.2	gallium
eka-silicon	72	72.0	germanium
eka-manganese	100	99	technetium (1939)
eka-molybdenum	140	not found	not found
eka-niobium	146	not found	not found
eka-cadmium	155	not found	not found
eka-iodine	170	not found	not found
eka-caesium	175	not found	not found
tri-manganese	190	186	rhenium (1925)
dvi-tellurium	212	210	polonium (1898)
dvi-caesium	220	223	francium (1939)
eka-tantalum	235	231	protactinium (1917)

21. Mendeleev's predictions, successful and otherwise

especially the case with Mendeleev and the periodic table, since he had to successfully accommodate as many as 63 known elements into a fully coherent system.

At the time of the discovery of the periodic table, the Nobel Prizes had not yet been instituted. One of the highest accolades in chemistry was the Davy medal, awarded by the Royal Society of Chemistry in Britain and named after the chemist Humphry Davy. In 1882, the Davy medal was jointly awarded to Lothar Meyer and Mendeleev. This fact seems to suggest that the chemists making the award were not over-impressed by Mendeleev's successful predictions, given that they were willing to recognize Lothar Meyer, who had not made any predictions to speak of. Moreover, the detailed citation that accompanied the award to Lothar Meyer and Mendeleev made absolutely no mention of Mendeleev's successful predictions. It would appear that at least this group of eminent British chemists were not swayed by the psychological impact of successful predictions over the ability of an individual to successfully accommodate the known elements.

The inert gases

The discovery of the inert gases at the end of the 19th century represented an interesting challenge to the periodic system for a number of reasons. First of all, in spite of Mendeleev's dramatic predictions of many other elements, he completely failed to predict this entire group of elements (He, Ne, Ar, Kr, Xe, Rn). Moreover, nobody else predicted these elements or even suspected their existence.

The first of them to be isolated was argon, in 1894, at University College in London. Unlike many previously discussed elements, a number of factors conspired together to render the accommodation of this element something of a Herculean task. It was not until six years later that the inert gases took their place as an eighth group between the halogens and the alkali metals.

But let us return to the first noble gas to be isolated, namely argon. It was obtained in small amounts, by Lord Rayleigh and William Ramsey who were working with the gas nitrogen. The all-important atomic weight of argon that was required if it were to be placed in the periodic table was not easily obtained. This was due to the fact that the atomicity of argon was itself not easy to determine. Most measurements pointed to its being monoatomic, but all the other known gases at the time were diatomic (H_2, N_2, O_2, F_2, Cl_2). If argon were indeed monoatomic its atomic weight would be approximately 40, which rendered its accommodation into the periodic table somewhat problematic, since there was no gap at this value of atomic weight. The element calcium has an atomic weight of about 40, followed by scandium, one of Mendeleev's successfully predicted elements, with an atomic weight of 44. This seemed to leave no space for a new element with an atomic weight of 40 (Figure 5).

There was a rather large gap between chlorine (35.5) and potassium (39), but placing argon between these two elements would have produced a rather obtrusive pair reversal. It should be recalled that at this time only one significant such pair reversal existed, involving the elements tellurium and iodine, and this behaviour was regarded as being highly anomalous. Mendeleev had concluded that the Te–I pair reversal was due to incorrect atomic weight determinations on either tellurium or iodine or perhaps both elements.

Yet another unusual aspect of the element argon was its complete chemical inertness, which meant that its compounds could not be studied, for the simple reason that it did not form any compounds. Some researchers took the inertness of the gas to mean that it was not a genuine chemical element. If this were true, it would provide an easy way out of the quandary of where to place the element as it would not need to be placed anywhere.

But many others persisted in attempting to place the element into the periodic table. The accommodation of argon was the focus of a general meeting of the Royal Society in 1885 and this was attended by the leading chemists and physicists of the day. The discoverers of argon, Rayleigh and Ramsey, argued that the element was probably monoatomic but admitted that they could not be sure. Nor could they be sure that the gas in question was not a mixture, which might imply that the atomic weight was not in fact 40. William Crookes presented some evidence in favour of sharp boiling and melting points of argon, thus pointing to a single element rather than a mixture. Henry Armstrong, a leading chemist, argued that argon might behave like nitrogen, in that it would form an inert diatomic molecule even though its individual atoms might be highly reactive.

A physicist, Arthur William Rücker, argued that an atomic weight of approximately 40 might be correct and that if this element could not be housed in the periodic table, then it was the periodic table itself that was at fault. This comment is interesting because it shows that even 16 years after the publication of Mendeleev's periodic table, and even after his three famous predictions had been fulfilled, not everyone was convinced of the validity of the periodic system.

The findings of the Royal Society meeting were therefore somewhat inconclusive about the fate of the new element argon, or whether indeed it should count as a new element. Mendeleev himself, who had not attended this meeting, published an article in the London-based *Nature* magazine in which he concluded that argon was in fact tri-atomic and that it consisted of three atoms of nitrogen. He based this conclusion on the fact that the assumed atomic weight of 40 divided by 3 gives approximately 13.3, which is not so far removed from 14, the atomic weight of nitrogen. In addition, argon had been discovered in the course of experiments on nitrogen gas, which rendered the tri-atomic idea more plausible.

The issue was finally resolved in the year 1900. At a conference in Berlin, Ramsey, one of the co-discoverers of the new gas, informed Mendeleev that the new group, which had by then been augmented with helium, neon, krypton, and xenon, could be elegantly accommodated in an eighth column between the halogens and the alkali metals. Argon, the first of these new elements, had been particularly troublesome because it represented a new case of a genuine pair reversal. It has an atomic weight of about 40 and yet appears before potassium with an atomic weight of about 39. Mendeleev now readily accepted this proposal and later wrote:

> This was extremely important for him [Ramsey] as an affirmation of the position of the newly discovered elements, and for me as a glorious confirmation of the general applicability of the periodic law.

Far from threatening the periodic table, the discovery of the inert gases and their successful accommodation into the periodic table only served to underline the great power and generality of Mendeleev's periodic system.

Chapter 6
Physics invades the periodic table

Although John Dalton had reintroduced the notion of atoms to science, many debates followed among chemists, most of whom refused to accept that atoms existed literally. One of these sceptical chemists was Mendeleev, but as we have seen in previous chapters, this does not seem to have prevented him from publishing the most successful periodic system of all those proposed at the time. Following the work of physicists like Einstein, the atom's reality became more and more firmly established, starting at the turn of the 20th century. Einstein's 1905 paper on Brownian motion, using statistical methods, provided conclusive theoretical justification for the existence of atoms but lacked experimental support. The latter was soon provided by the French experimental physicist Jean Perrin.

This change was accompanied by many lines of research aimed at exploring the structure of the atom, and developments which were to have a major influence on attempts to understand the periodic system theoretically. In this chapter, we consider some of this atomic research, as well as several other key discoveries in 20th-century physics that contributed to what I will call the invasion of the periodic table by physics.

22. Marie Curie

The discovery of the electron, the first subatomic particle and the first hint that the atom had a substructure, came in 1897 at the hands of the legendary J. J. Thomson at the Cavendish Laboratory in Cambridge. A little earlier, in 1895, Wilhelm Konrad Röntgen had discovered X-rays in Würzburg, Germany. These new rays would soon be put to very good use by Henry Moseley, a young physicist working first in Manchester and, for the remainder of his short scientific life, in Oxford.

Just a year after Röntgen had described his X-rays, Henri Becquerel in Paris discovered an enormously important phenomenon called radioactivity, whereby certain atoms were breaking up spontaneously while emitting a number of different new kinds of rays. The term 'radioactivity' was actually coined by the Polish-born Marie Sklodowska (later Curie) (Figure 22) who, with her husband Pierre Curie, took up the work on this dangerous new phenomenon and quickly discovered a couple of new elements that they called polonium and radium.

By studying how atoms break up while undergoing radioactive decay, it became possible to probe the components of the atom more effectively, as well as the laws that govern how atoms transform themselves into other atoms. So although the periodic table deals with distinct individuals or atoms of different elements, there seem to be some features that allow some atoms to be converted into others under the right conditions. For example, the loss of an alpha particle, consisting of a helium nucleus with two protons, results in a lowering of the atomic number of an element by two units.

Another very influential physicist working at this time was Ernest Rutherford, a New Zealander who arrived in Cambridge to pursue a research fellowship, later spending periods of time at McGill University in Montreal and at Manchester University. He then returned to Cambridge to assume the directorship of the Cavendish Laboratory as the successor to J. J. Thomson. Rutherford's contributions to atomic physics were many and varied, and included the discovery of the laws governing radioactive decay as well as being the first to 'split the atom'. He was also the first to achieve the 'transmutation' of elements into other new elements. In this way, Rutherford achieved an artificial analogue to the process of radioactivity which similarly yielded atoms of a completely different element and once again emphasized the essential unity of all forms of matter, another notion, incidentally, that Mendeleev had strenuously opposed during his lifetime.

Another discovery by Rutherford consisted of the nuclear model of the atom, a concept that is taken more or less for granted these days, namely the notion that the atom consists of a central nucleus surrounded by orbiting negatively charged electrons. His Cambridge predecessor, Thomson, had thought that the atom consisted of a sphere of positive charge within which the electrons circulated in the form of rings.

Nevertheless, Rutherford was not the first to suggest a nuclear model of the atom which resembled a miniature solar system. That distinction belongs to the French physicist Jean Perrin, who in 1900 proposed that negative electrons circulated around the positive nucleus like planets circulating around the Sun. In 1903, this astronomical analogy was given a new twist by Hantaro Nagaoka, from Japan, who proposed his Saturnian model in which the electrons now took the place of the famous rings around the planet Saturn. But neither Perrin nor Nagaoka could appeal to any solid experimental evidence to support their models of the atom, whereas Rutherford could.

Rutherford, along with his junior colleagues Geiger and Marsden, fired a stream of alpha particles at a thin metal foil made of gold and obtained a very surprising result. Whereas most of the alpha particles passed through the gold foil in a more or less unimpeded fashion, a significant number of them were deflected at very oblique angles. Rutherford's conclusion was that atoms of gold, or anything else for that matter, consist mostly of empty space except for a dense central nucleus. The fact that some of the alpha particles were unexpectedly bouncing back towards the incoming stream of alpha particles was thus evidence for the presence of a tiny central nucleus in every atom.

Nature therefore turned out to be more fluid than had previously been believed. Mendeleev, for example, had thought that elements were strictly individual. He could not accept the notion that elements could be converted into different ones. In fact, after the Curies began to report experiments that suggested the breaking up of atoms, Mendeleev travelled to Paris to see the evidence for himself, close to the end of his life. It is not clear whether he accepted this radical new notion even after his visit to the Curie laboratory.

X-rays

In 1895, the German physicist Röntgen made a momentous discovery at the age of 40. Up to this point, his research output had been somewhat unexceptional. As the atomic physicist Emilio Segrè wrote a good deal later:

> By the beginning of 1895 Röntgen had written forty-eight papers now practically forgotten. With his forty-ninth he struck gold.

Röntgen was in the process of exploring the action of an electric current in an evacuated glass tube called a Crookes tube. Röntgen noticed that an object, which was not part of his experiment, on the other side of his lab was glowing. It was a screen coated with barium platinocyanide. He quickly established that the glow was not due to the action of the electric current and deduced that some new form of rays might be produced inside his Crookes tube. Soon, Röntgen also discovered the property that X-rays are best known for. He found that he could produce an image of his hand that clearly showed the outlines of just his bones. A powerful new technique that was to provide numerous medical applications had come to light. After working secretly for seven weeks, Röntgen was ready to announce his results to the Würzburg Physical-Medical Society, an interesting coincidence given the impact that his new rays were to have on these two fields that were still remote from each other at the time.

Some of Röntgen's original X-ray images were sent to Paris, where they reached Henri Becquerel, who became interested in examining the relationship between X-rays and the property of phosphorescence, whereby certain substances emit light on exposure to sunlight. In order to test this notion, Becquerel wrapped some crystals of a uranium salt in some thick paper and, because of a lack of sunlight, decided to put these materials away into a drawer for a few days. By another stroke of luck, Becquerel happened to place his wrapped crystals on top of an undeveloped photographic plate before closing the drawer and going about his business for a few more cloudy Parisian days.

On finally opening the drawer, he was amazed to find an image that had been created on the photographic plate by the uranium crystals even though no sunlight had struck them. This clearly suggested that the uranium salt was emitting its own rays, irrespective of the process of phosphorescence. Becquerel had discovered nothing less than radioactivity, a natural process in some materials whereby the spontaneous decay of atoms produces powerful and, in some cases, dangerous emanations. It was Marie Curie who was to dub this phenomenon 'radioactivity' a few years later.

The supposed connection between these experiments and X-rays turned out to be incorrect. Becquerel failed to find any connection between X-rays and phosphorescence. In fact, X-rays had not even entered into these experiments, and yet he discovered a phenomenon that was to have enormous importance in more ways than one. First, radioactivity was an early and significant step into the exploration of matter and radiation, and secondly, it was to lead indirectly to the development of nuclear weapons.

Back to Rutherford

Around the year 1911, Rutherford reached the conclusion that the charge on the nucleus of the atom was approximately half of the weight of the atom concerned, or $Z \approx A/2$, after analysing the results of atomic scattering experiments. This conclusion was supported by the Oxford physicist Charles Barkla, who arrived at it via an altogether different route, using experiments with X-rays.

Meanwhile, a complete outsider to the field, the Dutch econometrician Anton van den Broek, was pondering over the possibility of modifying Mendeleev's periodic table. In 1907, he proposed a new table containing 120 elements, although many of these remained as empty spaces. A good number of empty spaces were occupied by some newly discovered substances whose elemental status was still in some question. They included so-called thorium emanation, uranium-X (an unknown decay product of uranium), Gd_2 (a decay product from gadolinium), and many other new species.

But the really novel feature of van den Broek's work was a proposal that all elements were composites of a particle that he named the 'alphon', consisting of half of a helium atom with a mass of two atomic weight units. In 1911, he published a further article in which he dropped any mention of alphons, but retained the idea of elements differing by two units of atomic weight. In a 20-line letter to London's *Nature* magazine, he went a step further towards the concept of atomic number by writing,

> ...the number of possible elements is equal to the number of possible permanent charges...

Van den Broek was thus suggesting that since the nuclear charge on an atom was half of its atomic weight, and that the atomic weights of successive elements increased in step-wise fashion by two, then the nuclear charge defined the position of an element in the periodic table. In other words, each successive element in the periodic table would have a nuclear charge greater by one than the previous element.

A further article published in 1913 drew the attention of Niels Bohr, who cited van den Broek in his own famous trilogy paper of 1913 on the hydrogen atom and the electronic configurations of many-electron atoms. In the same year, van den Broek wrote another paper which also appeared in *Nature*, this time explicitly connecting the serial number on each atom with the charge on each atom. More significantly perhaps, he disconnected this serial number from atomic weight. This landmark publication was praised by many experts in the field, including Soddy and Rutherford, all of whom had failed to see the situation as clearly as the amateur van den Broek.

Moseley

Although an amateur had confounded the experts in arriving at the concept of atomic number, he did not quite complete the task of establishing this new quantity. The person who did complete it, and who is almost invariably given the credit for the discovery of atomic number, was the English physicist Henry Moseley, who died in the First World War at the age of 26. His fame rests on just two papers in which he confirmed experimentally that atomic number was a better ordering principle for the elements than atomic weight. This research is also important because it allowed others to determine just how many elements were still awaiting discovery between the limits of the naturally occurring elements (hydrogen and uranium).

Moseley received his training at the University of Manchester as a student of Rutherford's. Moseley's experiments consisted of bouncing light off the surface of samples of various elements and recording the characteristic X-ray frequency that each one emitted. Such emissions occur because an inner electron is ejected from the atom, causing an outer electron to fill the empty space, in a process which is accompanied by the emission of X-rays.

Moseley first selected 14 elements, 9 of which, titanium to zinc, formed a continuous sequence of elements in the periodic table. What he discovered was that a plot of the emitted X-ray frequency against the square of an integer representing the position of each element in the periodic table produced a straight-line graph. Here was confirmation of van den Broek's hypothesis that the elements can be ordered by means of a sequence of integers, later called atomic number, one for each element starting with H = 1, He = 2, and so on.

In a second paper, he extended this relationship to a further 30 elements, thus solidifying its status further. It then became a relatively simple matter for Moseley to verify whether the claims for many newly claimed elements were valid or not. For example, the Japanese chemist Seiji Ogawa had claimed that he had isolated an element to fill the space below manganese in the periodic table. Moseley measured the frequency of X-rays that Ogawa's sample produced when bombarded with electrons and found that it did not correspond to the value expected of element 43.

While chemists had been using atomic weights to order the elements there had been a great deal of uncertainty about just how many elements remained to be discovered. This was due to the irregular gaps that occurred between the values of the atomic weights of successive elements in the periodic table. This complication disappeared when the switch was made to using atomic number. Now the gaps between successive elements became perfectly regular, namely one unit of atomic number.

After Moseley had died, other chemists and physicists used his method and found that the remaining unknown elements were those with atomic numbers of 43, 61, 72, 75, 85, 87, and 91 among the evenly spaced sequence of atomic numbers. It was not until 1945 when the last of these gaps was filled, following the synthesis of element number 61, or promethium.

Isotopes

The discovery of isotopes of any particular element is another key step in understanding the periodic table which occurred at the dawn of atomic physics. The term comes from *iso* (same) and *topos* (place) and is used to describe atomic species of any particular element which differ in weight and yet occupy the same place in the periodic table. The discovery came about partly as a matter of necessity. The new developments in atomic physics led to the discovery of a number of new elements such as Ra, Po, Rn, and Ac which easily assumed their rightful places in the periodic table. But in addition, 30 or so more apparent new elements were discovered over a short period of time. These new species were given provisional names like thorium emanation, radium emanation, actinium X, uranium X, thorium X, and so on, to indicate the elements which seemed to be producing them. The Xs denoted unknown species, which turned out to be isotopes of different elements in most cases. For example, uranium X was later recognized to be an isotope of thorium.

Some designers of periodic tables, like van den Broek, attempted to accommodate these new 'elements' into extended periodic tables as we saw above. Meanwhile, two Swedes, Daniel Strömholm and Theodor Svedberg, produced periodic tables in which some of these exotic new species were forced into the same place. For example, below the inert gas xenon, they placed radium emanation, actinium emanation, and thorium emanation. This seems to represent an anticipation of isotopy, but not a very clear recognition of the phenomenon.

In 1907, the year of Mendeleev's death, the American radiochemist Herbert McCoy concluded that 'radiothorium is entirely inseparable from thorium by chemical processes'. This was a key observation which was soon repeated in the case of many other pairs of substances that had originally been thought to be new elements. The full appreciation of such observations was made by Frederick Soddy, yet another former student of Rutherford.

To Soddy, the chemical inseparability meant only one thing, namely that these were two forms, or more, of the same chemical element. In 1913, he coined the term 'isotopes' to signify two or more atoms of the same element which were chemically completely inseparable, but which had different atomic weights. Chemical inseparability was also observed by Friedrich Paneth and Georg von Hevesy in the case of lead and 'radio lead', after

Rutherford had asked them to separate them chemically. After attempting this feat by 20 different chemical approaches, they were forced to admit complete failure. It was a failure in some respects but one that solidified further the notion of one element, lead in this case, occurring as chemically inseparable isotopes. But it was not all wasted work because Paneth and von Hevesy's efforts allowed them to establish a new technique for the radioactive labelling of molecules, which became the basis for a very useful subdiscipline with far-reaching applications, in areas like biochemistry and medical research.

In 1914, the case for isotopy gained even greater support from the work of T. W. Richards at Harvard, who set about measuring the atomic weights of two isotopes of the same element. He too selected lead, since this element was produced by a number of radioactive decay series. Not surprisingly, the lead atoms formed by these alternative pathways, involving quite different intermediate elements, resulted in the formation of atoms of lead differing by a large value of 0.75 of an atomic weight unit. This result was later extended by others to 0.85 of a unit.

Finally, the discovery of isotopes further clarified the occurrence of pair reversals, such as in the case of tellurium and iodine that had plagued Mendeleev. Tellurium has a higher atomic weight than iodine, even though it precedes it in the periodic table, because the weighted average of all the isotopes of tellurium happens to be a higher value than the weighted average of the isotopes of iodine. Atomic weight is thus seen as a contingent quantity depending upon the relative abundance of all the isotopes of an element. The more fundamental quantity, as far as the periodic table is concerned, is the van den Broek–Moseley atomic number or, as it was later realized, the number of protons in the nucleus. The identity of an element is captured by its atomic number and not its atomic weight, since the latter differs according to the particular sample from which the element has been isolated.

Whereas tellurium has a higher average atomic weight than iodine, its atomic number is one unit smaller. If one uses atomic number instead of atomic weight as an ordering principle for the elements, both tellurium and iodine fall into their appropriate groups in terms of chemical behaviour. It emerged, therefore, that pair reversals had only been required because an incorrect ordering principle had been used in all periodic tables prior to the turn of the 20th century.

Chapter 7
Electronic structure

The last chapter dealt mostly with discoveries in classical physics that did not require quantum theory. This was true of X-rays and radioactivity, which were largely studied without any quantum concepts, although the theory was later used to clarify certain aspects. In addition, the physics described in the last chapter was mostly about processes originating in the nucleus of the atom. Radioactivity is essentially about the break-up of the nucleus, and the transmutation of elements likewise takes place at the level of the nucleus. Moreover, atomic number is a property of the nuclei of atoms, and isotopes are distinguished by different masses of atoms of the same element, which are made up almost exclusively by the mass of their nuclei.

In this chapter, we venture into discoveries concerning the electrons in atoms, a study that did necessitate the use of quantum theory right from its early days. But first, we will need to say something of the origins of quantum theory itself. It all began before the turn of the 20th century in Germany, where a number of physicists were attempting to understand the behaviour of radiation held in a small cavity with blackened walls. The spectral behaviour of such 'black body radiation' was carefully recorded at different temperatures and attempts were then made to model the resulting graphs mathematically. The problem remained unsolved for a good deal of time until Max Planck made the bold assumption that the energy of this radiation consisted of discrete packets or 'quanta' in the year 1900. Planck himself seems to have been reluctant to accept the full significance of his new quantum theory and it was left to others to make some new applications of it.

The quantum theory asserted that energy comes in discrete bundles and that no intermediate values can occur between certain whole number multiples of the basic quantum of energy. This theory was successfully applied to the photo-electric effect in 1905 by none other than Albert Einstein, perhaps the most brilliant physicist of the 20th century. The outcome of his research was that light could be regarded as having a quantized, or particulate, nature. Nevertheless, Einstein soon began to regard quantum mechanics as an incomplete theory and was to maintain his criticism of it for the remainder of his life.

In 1913, the Dane Niels Bohr applied quantum theory to the hydrogen atom that he supposed, like Rutherford, to consist of a central nucleus with a circulating electron. Bohr assumed that the energy available to the electron occurred only in certain discrete values or, in pictorial terms, that the electron could exist in any number of shells or orbits around the nucleus. This model could explain, to some extent, a couple of features of the behaviour of the hydrogen atom and in fact atoms of any element. First, it explained why atoms of hydrogen which were exposed to a burst of electrical energy would result in a discontinuous spectrum in which only some rather specific frequencies were observed. Bohr reasoned that such behaviour came about when an electron underwent a transition from one allowed energy level to another one. Such a transition was accompanied by the release, or absorption, of the precise energy corresponding to the energy difference between the two energy levels in the atom.

Secondly, and less satisfactorily, the model explained why electrons did not lose energy and crash into the nucleus of any atom, as predicted from applying classical mechanics to a charge particle undergoing circular motion. Bohr's response was that the electrons would simply not lose energy provided that they remained in their fixed orbits. He also postulated that there was a lowest energy level beyond which the electron could not undergo any downward transitions.

Bohr then generalized his model to cover any many-electron atom rather than just hydrogen. He also set about trying to establish the way in which the electrons were arranged in any particular atom. Whereas the theoretical validity of making such a leap from one-electron to many-electrons was in question, this did not deter Bohr from forging ahead regardless. The electronic configurations that he arrived at are shown in Figure 23.

But Bohr's assignment of electrons to shells was not carried out on mathematical grounds, nor with any explicit help from the quantum theory. Instead, Bohr appealed to chemical evidence such as the knowledge that atoms of the element boron can form three bonds, as do other elements in the boron group. An atom of boron therefore had to have three outer-shell electrons in order for this to be possible. But even with such a rudimentary and non-deductive theory, Bohr was providing the first successful electron-based explanation for why such elements as lithium, sodium, and potassium occur in the same group of the periodic table, and likewise for the membership of any group in the periodic table. In the case of lithium, sodium, and potassium, it is because each of these atoms has one electron that is set aside from the remaining electrons in an outer shell.

Bohr's theory had some other limitations, one of them being that it was only strictly applicable to one-electron atoms such as hydrogen or ions like He^+, Li^{2+}, Be^{3+} etc. It was also found that some of the lines in the spectra for these 'hydrogenic' spectra broke up into unexpected pairs of lines. Arnold Sommerfeld working in Germany suggested that the nucleus might lie at one of the foci

1	H	1				
2	He	2				
3	Li	2	1			
4	Be	2	2			
5	B	2	3			
6	C	2	4			
7	N	4	3			
8	O	4	2	2		
9	F	4	4	1		
10	Ne	8	2			
11	Na	8	2	1		
12	Mg	8	2	2		
13	Al	8	2	3		
14	Si	8	2	4		
15	P	8	4	3		
16	S	8	4	2	2	
17	Cl	8	4	4	1	
18	Ar	8	8	2		
19	K	8	8	2	1	
20	Ca	8	8	2	2	
21	Sc	8	8	2	3	
22	Ti	8	8	2	4	
23	V	8	8	4	3	
24	Cr	8	8	2	2	2

23. **Bohr's original 1913 scheme for electronic configurations of atoms**

of an ellipse rather than at the heart of a circular atom. His calculations showed that one had to introduce subshells within Bohr's main shells of electrons. Whereas Bohr's model was characterized by one quantum number denoting each of the separate shells or orbits, Sommerfeld's modified model required two quantum numbers to specify the elliptical path of the electron. Armed with these new quantum numbers Bohr was able to compile a more detailed set of electronic configurations in 1923, as shown in Figure 24.

H	1				
He	2				
Li	2	1			
Be	2	2			
B	2	3			
C	2	4			
N	2	4	1		
O	2	4	2		
F	2	4	3		
Ne	2	4	4		
Na	2	4	4	1	
Mg	2	4	4	2	
Al	2	4	4	2	1
Si	2	4	4	4	
P	2	4	4	4	1
S	2	4	4	4	2
Cl	2	4	4	4	3
Ar	2	4	4	4	4

24. Bohr's 1923 electronic configurations based on two quantum numbers

A few years later the English physicist Edmund Stoner found that a third quantum number was also needed to specify some finer details of the spectrum of hydrogen and other atoms. Then in 1924 the Austrian-born theorist, Wolfgang Pauli, discovered the need for a fourth quantum number, which was identified with the concept of an electron adopting one of two values of a special kind of angular momentum. This new kind of motion was eventually called electron 'spin' even though electrons do not literally spin in the same way that the earth spins about an axis while also performing an orbital motion around the sun.

The four quantum numbers are related to each other by a set of nested relationships. The third quantum number depends on the value of the second quantum number that in turn depends on that of the first quantum number. Pauli's fourth quantum number is a little different since it can adopt values of +1/2 or –1/2 regardless of the values of the other three quantum numbers. The importance of the fourth quantum number, especially, is that its arrival provided a good explanation for why each electron shell can contain a certain number of electrons (2, 8, 18, 32 etc.), starting with the shell closest to the nucleus.

Here is how this scheme works. The first quantum number n can adopt any integral value starting with 1. The second quantum number, which is given the label ℓ, can have any of the following values related to the values of n,

$$\ell = n - 1, \ldots \ldots 0$$

In the case when n = 3, for example, ℓ can take the values 2, 1, or 0. The third quantum number labelled ml can adopt values related to those of the second quantum numbers as,

$$m_\ell = -\ell, -(\ell + 1), \ldots 0 \ldots (\ell - 1), \ell$$

For example, if $\ell = 2$, the possible values of m_ℓ are,

$$-2, -1, 0, +1, +2$$

Finally, the fourth quantum number labelled m_s can only take two possible values, either $+1/2$ or $-1/2$ units of spin angular momentum as mentioned before. There is therefore a hierarchy of related values for the four quantum numbers, which are used to describe any particular electron in an atom (Figure 25).

As a result of this scheme it is clear why the third shell, for example, can contain a total of 18 electrons. If the first quantum number, given by the shell number, is 3, there will be a total of $2 \times (3)^2$ or 18 electrons in the third shell. The second quantum number ℓ can take values of 2, 1, or 0. Each of these values of ℓ will generate a number of possible values of m_ℓ and each of these values will be multiplied by a factor of two since the fourth quantum number can adopt values of $1/2$ or $-1/2$.

n	Possible values of ℓ	Subshell Designation	Possible Values of m_ℓ	Orbitals in Subshell	Electrons in each shell
1	0	1s	0	1	2
2	0	2s	0	1	
	1	2p	1, 0, −1	3	8
3	0	3s	0	1	
	1	3p	1, 0, −1	3	
	2	3d	2, 1, 0, −1, −2	5	18
4	0	4s	0	1	
	1	4p	1, 0, −1	3	
	2	4d	2, 1, 0, −1, −2	5	
	3	4f	3, 2, 1, 0, −1, −2, −3	7	32

25. Combination of four quantum numbers to explain the total number of electrons in each shell

But the fact that the third shell can contain 18 electrons does not strictly explain why it is that some of the periods in the periodic system contain 18 places. It would only be a rigorous explanation of this fact if electron shells were filled in a strictly sequential manner. Although electron shells begin by filling in a sequential manner, this ceases to be the case starting with element number 19 or potassium. Configurations are built up starting with the 1s orbital which can contain 2 electrons, moving to the 2s electron which likewise is filled with another 2 electrons. Then come the 2p orbitals which altogether contain a further 6 electrons, and so on. This process continues in a predictable manner up to element 18 or argon which has the configuration of

$$1s^2, 2s^2, 2p^6, 3s^2, 3p^6$$

One might expect that the configuration for the subsequent element, number 19, or potassium, would be

$$1s^2, 2s^2, 2p^6, 3s^2, 3p^6, 3d^1$$

where the final electron occupies the next subshell, which is labelled as 3d. This would be expected because up to this point the

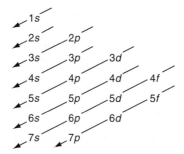

26. Order of filling of atomic orbitals. Follow diagonal arrows from top to bottom

pattern has been one of adding the differentiating electron to the next available orbital at increasing distances from the nucleus. However, experimental evidence shows that the configuration of potassium should be denoted as

$$1s^2, 2s^2, 2p^6, 3s^2, 3p^6, 4s^1$$

Similarly, in the case of element 20, or calcium, the new electron also enters the 4s orbital. But in the next element, number 21 or scandium, the configuration is observed to be

$$1s^2, 2s^2, 2p^6, 3s^2, 3p^6, 4s^2, 3d^1$$

This kind of skipping backwards and forwards among available orbitals as the electrons fill successive elements recurs again several times. The order of filling is summarized in Figure 26 above.

As a consequence of this order of filling, successive periods in the periodic table contain the following number of elements, 2, 8, 8, 18, 18, 32, etc., thus showing a 'doubling' for each period except the first one.

Whereas the rule for the combination of four quantum numbers provides a rigorous explanation for the point at which shells close, it does not provide an equally rigorous explanation for the point at which periods close. Nevertheless, some rationalizations can be given for this order of filling, but such rationalizations tend to be somewhat dependent on the facts one is trying to explain. We know where the periods close because we know that the noble gases occur at elements 2, 10, 18, 36, 54, etc. Similarly, we have a knowledge of the order of orbital filling from observations but not from theory. The conclusion, seldom acknowledged in textbook accounts of the explanation of the periodic table, is that quantum physics only partly explains the periodic table. Nobody has yet deduced the order of orbital filling from the principles of quantum mechanics. This is not to say that it might not be achieved in the future or that the order of electron filling is in any sense inherently irreducible to quantum physics.

Chemists and configurations

The discovery of the electron by J. J. Thomson in 1897 opened up all kinds of new explanations in physics as well as entirely new lines of experimentation. Thomson was also one of the first to discuss the way in which the electrons are arranged in atoms, although his theory was not very successful because not enough was known about how many electrons existed in any particular atom. As we have seen, the first significant theory of this kind was due to Bohr who also introduced the notion of quantization of energy into the realm of the atom and into the assignment of electronic arrangements. Bohr succeeded in publishing a set of electronic configurations for many of the known atoms but only after consulting the chemical and spectral behaviour that had been gathered together by others over many years.

But what were the chemists doing at about this time? How did their attempts to exploit the electron compare with those of Bohr and other quantum physicists? To begin this survey requires going back in time to 1902 just after the electron had been discovered. The American chemist G. N. Lewis, who was working in the Philippines

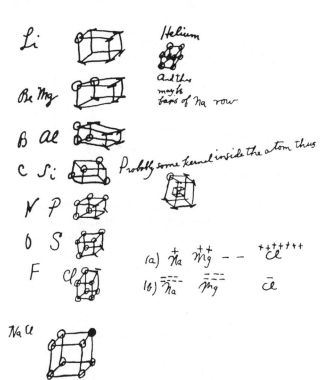

27. Lewis's sketch of atomic cubes

at the time, sketched a diagram that is shown in Figure 27, the original copy of which still exists to this day. In it, he assumed that electrons occur at the corners of a cube and that as we move through the periodic table, one element at a time, a further electron is added to each corner. The choice of a cube may seem to be rather bizarre from a modern perspective since we now know that electrons circulate around the nucleus. But Lewis's model makes good sense in one important respect that is connected with the periodic table. Eight is the number of elements through which one must progress before a repetition in properties occurs.

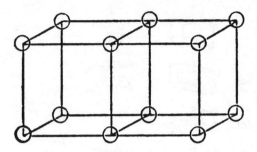

28. Lewis's depiction of a double bond between two atoms

Lewis was therefore suggesting that chemical periodicity and the properties of individual elements were governed by the number of electrons in the outermost electron cube around the nucleus of an atom. All that is wrong with the model is that the electrons are regarded as being stationary, but the choice of a cube is a natural and ingenious one which reflects the fact that chemical periodicity is based on intervals of eight elements.

In the same famous sketch, Lewis gave a diagram of how sodium and chlorine atoms can form a compound, after an electron is transferred from a sodium atom to one of chlorine, in order to occupy the place of the missing eighth electron in the outer cube of chlorine. Lewis then waited for a period of fourteen years before publishing these ideas as well as extending them to include another form of bonding, namely covalent bonding, that involves the sharing rather than the transfer of electrons between various atoms.

By considering many known compounds, and by counting the number of outer electrons that their atoms possessed, Lewis arrived at the conclusion that in most cases the electron count produced an even number. This fact suggested to him that chemical bonding might be due to the pairing of electrons, an idea that soon became central to all of chemistry and which remains essentially correct to this day, even following the subsequent quantum mechanical theories of chemical bonding.

1	2	3	4	5	6	7
H						
Li	Be	B	C	N	O	F
Na	Mg	Al	Si	P	S	Cl
K	Ca	Sc		As	Se	Br
Rb	Sr			Sb	Te	I
Cs	Ba			Bi		

29. Lewis's outer electronic structures for 29 elements. The number at the head of the column represents the positive charge on the atomic kernel and also the number of outer shell electrons for each atom (compiled by the author)

In order to represent the sharing of electrons between two atoms, Lewis drew two adjacent cubes which share an edge, or two electrons. Similarly, a double bond was represented by two cubes that share a common face or four electrons (Figure 28). But then a problem arose. In organic chemistry, it was known that some compounds, such as acetylene or C_2H_2, contain a triple bond. Lewis realized that his model of electrons on the corners of a cube could not represent such triple bonds. In the same paper, he switched to a new model in which four pairs of electrons were located at the corners of a tetrahedron instead of a cube. A triple bond was then rendered by two tetrahedra that share a common face.

In the same paper, Lewis also returned to the question of electronic configurations of atoms and now extended his arrangements to include 29 elements as shown in Figure 29. Before moving on to other contributors it is worth pausing to consider that G. N. Lewis is possibly the most significant chemist of the 20th century not to have been awarded a Nobel Prize. This partly stems from his having died an early death in his own laboratory as a result of hydrogen cyanide poisoning and the fact that Nobel Prizes are not awarded posthumously. Last but not

		TABLE I.										
Classification of the Elements According to the Arrangement of Their Electrons.												
Layer.	N	E = o	1	2	3	4	5	6	7	8	9	10
I			H	He								
IIa	2	He	Li	Be	B	C	N	O	F	Ne		
IIb	10	Ne	Na	Mg	Al	Si	P	S	Cl	A		
IIIa	18	A	K	Ca	Sc	Ti	V	Cr	Mn	Fe	Co	Ni
			11	12	13	14	15	16	17	18		
IIIa	28	Niβ	Cu	Zn	Ga	Ge	As	Se	Br	Kr		
IIIb	36	Kr	Rb	Sr	Y	Zr	Cb	Mo	43	Ru	Rh	Pd
			11	12	13	14	15	16	17	18		
IIIb	46	Pdβ	Ag	Cd	In	Sn	Sb	Te	I	Xe		
IVa	54	Xe	Cs	Ba	La	Ce	Pr	Nd	61	Sa	Eu	Gd
			11	12	13	14	15	16	17	18		
IVa			Tb	Ho	Dy	Er	Tm	Tm$_2$	Yb	Lu		
	14		15	16	17	18	19	20	21	22	23	24
IVa	68	Erβ	Tmβ	Tm$_2$β	Ybβ	Luβ	Ta	W	75	Os	Ir	Pt
			25	26	27	28	29	30	31	32		
IVa	78	Ptβ	Au	Hg	Tl	Pb	Bi	RaF	85	Nt		
IVb	86	Nt	87	Ra	Ac	Th	Ux$_2$	U				

30. Langmuir's periodic table

least, Lewis made too many enemies in the course of his academic career and so did not endear himself to colleagues who might otherwise have nominated him for the award.

Lewis's ideas were extended and popularized by the American industrial chemist Irving Langmuir. Whereas Lewis had assigned electronic configurations to only 29 elements, Langmuir set out to complete the task. While Lewis had refrained from assigning configurations to the atoms of transition metal elements, Langmuir gave the following list in a paper dated 1919,

Sc	Ti	V	Cr	Mn	Fe	Co	Ni	Cu	Zn
3	4	5	6	7	8	9	10	11	12

As Lewis had done before him, Langmuir used the chemical properties of the elements to guide him into these assignments and not any arguments from quantum theory (Figure 30). Not surprisingly, these chemists were able to improve upon the electronic configurations which physicists like Bohr were groping towards.

In 1921, the English chemist Charles Bury, working at the University College of Wales in Aberystwyth, challenged an idea that was implicit in the work of both Lewis and Langmuir. This is the assumption that electron shells are filled sequentially as one moves through the periodic table adding an extra electron for each new element encountered, as already mentioned on page 86. Bury claimed that his own refined set of electronic configurations provided better agreement with the known chemical facts.

> Since eight is the maximum number of electrons in an outer layer K, Ca and Sc must form a fourth layer although their third is not complete. Their structures will be 2, 8, 8, 1, 2, 8, 8, 2, 2, 8, 8, 3.

Bury also broke with Lewis and Langmuir by suggesting that inner and apparently stable groups of 8 electrons could change into groups of 18 electrons and similarly that groups of 18 electrons could change into groups of 32. These features represent some early signs of the emergence of the medium-long- and long-form tables respectively.

To conclude, physicists provided a great impetus to attempts to understand the underlying basis of the periodic table but the chemists of the day were in many cases able to apply the new physical ideas, such as electronic configurations, to better effect.

Chapter 8
Quantum mechanics

The previous chapter considered the influence of early quantum theory, especially that of Bohr, on the explanation of the periodic table. This explanation culminated in Pauli's contribution, his introduction of a fourth quantum number and the Exclusion Principle that now bears his name. It then became possible to explain why each shell around the atom can contain a particular number of electrons (first shell – 2, second shell – 8, third shell – 18, etc.). If one assumes the correct order of filling of orbitals within these shells one can now explain the fact that the period lengths are actually, 2, 8, 8, 18, 18, etc. But any worthy explanation of the periodic table should be capable of deriving this sequence of values from first principles, *without* assuming the observed order of orbital filling.

The quantum theory of Bohr, even when augmented by Pauli's contributions, was therefore only a stepping stone to a more advanced theory. The Bohr–Pauli version is usually called quantum theory or sometimes the old quantum theory to distinguish it from the later development which was to take place in 1925 and 1926 which became known as quantum mechanics. The use of the word 'theory' to denote the older version is a little unfortunate because it reinforces the common misconception that a theory is somewhat vague and on its way to becoming a scientific law or some other more solid piece of knowledge.

But in science, a theory is a highly supported, although never proven, body of knowledge which if anything possesses a more elevated status than scientific laws. Many different laws are often subsumed within one over-arching 'theory'. So quantum mechanics, the successor theory, was every bit as much a 'theory' as was Bohr's old quantum theory, even though it was a more general and more successful theory.

Bohr's old quantum theory had several shortcomings including the fact that it could not explain chemical bonding, but this all changed following the advent of quantum mechanics. It then became possible to go beyond the picaresque idea of G. N. Lewis that bonding resulted from the mere sharing of electrons between two or more atoms in a molecule. According to quantum mechanics, electrons behave as much as waves as they do as particles. By writing a wave equation for the motion of electrons around the nucleus, the Austrian Erwin Schrödinger was able to take a decisive step forward. The solutions of Schrödinger's equation, of which there are many, represent the possible quantum states that the electrons in an atom can find themselves in. A little later two physicists, Hund and Mulliken, independently developed the molecular orbital theory in which bonding was found to occur as a result of constructive and destructive interference between electron waves on each atom in a molecule.

But we need to return to atoms and the periodic table. Using Bohr's theory the only atoms whose energy levels could be calculated were those consisting of just one electron, which includes the H atom and one-electron ions such as He^{+1}, Li^{+2}, Be^{+3}, etc. In the case of many-electron atoms, where many means more than one, Bohr's old quantum theory was impotent. Conversely, using the new quantum mechanics many-electron atoms could be handled by theoreticians, although admittedly in an approximate fashion instead of exactly. This is due to a mathematical limitation. Any system with more than one electron consists of a so-called 'many-body problem', and such problems only admit approximate solutions.

So the energies of the quantum states for any many-electron atom can be approximately calculated from first principles although there is extremely good agreement with observed energy values. Nevertheless, some global aspects of the periodic table have still not been derived from first principles to this day. For example, the order of orbital filling that was mentioned earlier is a case in point.

As we saw in the previous chapter, atomic orbitals which make up subshells and shells in any atom are filled in order of increasing values of the sum of the first two quantum numbers denoting any particular orbital. This fact is summarized by the $n + \ell$ or Madelung rule. Orbitals fill in such a way that the quantity $n + \ell$ increases gradually, starting with a value of 1 for the 1s orbital.

By following the diagonal arrows starting at the 1s orbital, and then by moving to the next diagonal set of lines, one obtains the order of orbital filling, which is as follows (Figure 26),

$$1s < 2s < 2p < 3s < 3p < 4s < 3d < 4p < 5s \text{ etc.}$$

In spite of its many dazzling successes, quantum mechanics has still not succeeded in deriving this sequence of $n + \ell$ values in a purely theoretical fashion. This is not to diminish the successes of the newer theory which are truly impressive. It is just to point out that a feature remains to be elucidated. Perhaps a quantum physicist may succeed in deriving the Madelung rule in the future, or perhaps it may require an even more powerful future theory to do so. I am not trying to suggest some kind of inherent and 'spooky' irreducibility of chemical phenomena to quantum physics but am

31. Niels Bohr

just concentrating on what has actually been achieved so far in the context of the periodic table.

But let's turn to what quantum mechanics has illuminated regarding chemical periodicity. As was mentioned earlier, the Schrödinger equation can be written down and solved for any atom system in the periodic table without any experimental input whatsoever. For example, physicists and theoretical chemists have succeeded in calculating the ionization energies of the atoms in the periodic table. When these calculations are compared with experimental values a remarkable degree of agreement is found (Figure 32).

Now it so happens that ionization energy is one of many properties of atoms which show marked periodicity. As we increase the atomic number starting with Z = 1 for hydrogen, the ionization energy shows an increase until we reach the very next element, helium. This is followed by a sharp decrease before we arrive at Z = 3 or lithium. Subsequent values of ionization energy are then found to increase, broadly speaking, until we reach an element that is chemically analogous to helium, namely neon, which is another noble gas. This pattern of overall increase in ionization energy then recurs across all periods of the periodic table. And for each period the minimum value occurs for elements in group 1 such as lithium, sodium, and potassium. Maximum values, as we have just seen, occur at the noble gases such as helium, neon, argon, krypton, and xenon. Figure 32 also shows the curve obtained when theoretically computed values are connected together.

The conclusion is that even if quantum mechanics has not yet produced a general derivation of the equation for the order of orbital filling (the n + ℓ rule), it has still explained the periodicity in the properties of the atoms of all elements, although this is carried out on a one-at-a-time basis for each separate atom rather than through a universal one-for-all solution. How has this been possible in quantum mechanics as compared with Bohr's old quantum theory?

In order to answer this question, we need to delve a little deeper into the difference between the two theories. Perhaps the best place to begin is with a brief discussion of the nature of waves. Many phenomena in physics present themselves in the form of waves. Light is propagated as light waves and sound as sound waves. When a rock is thrown into a pond, a series of ripples or water waves begin to radiate from the point at which the rock has

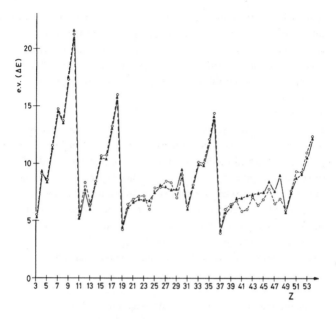

32. Theoretically calculated (triangles) and observed ionization energies (circles) of elements 1 to 53

entered the water. Two interesting phenomena are associated with waves of these three kinds or indeed with waves of any kind. First of all, waves display an effect called diffraction, which describes the way in which they tend to spread when they pass through an aperture or around any obstacle.

In addition, two or more waves arriving at a screen together are said to be 'in-phase' and produce an effect called constructive interference which leads to an overall increase in intensity over and above the intensities of the two separate waves. Conversely, two waves arriving at a screen in an out of phase fashion bring about destructive interference, the result of which is a cancellation of wave intensity. In the early 1920s, for reasons that we cannot go into here, it was suspected that particles like electrons might just behave as waves under certain conditions. Briefly put this idea was a logical inversion of Einstein's work on the photoelectric effect, in which it had been established that light waves also behave as particles.

In order to test whether particles such as electrons really behave as waves, it became necessary to see whether electrons might also produce diffraction and interference effects like all other kinds of waves can. Surprising as it might seem, such experiments succeeded and the wave nature of electrons was established once and for all. In addition, a beam of electrons directed at a single crystal of metallic nickel was found to produce a series of concentric rings, that is to say a diffraction pattern was formed due to electron waves spreading around the crystal rather than just bouncing off it.

From this point onwards, electrons, and other fundamental particles, had to be regarded as having a kind of schizophrenic nature, behaving as both particles and waves. More specifically, the news that electrons behave as waves quickly reached the wider community of theoretical physicists, including one Erwin Schrödinger. This Austrian-born theorist set about using the well-trodden mathematical techniques for modelling waves in order to describe the behaviour of the electron in a hydrogen atom. His assumptions were first that electrons behaved as waves and secondly that the potential energy of the electron consisted of the energy with which it was attracted by the nucleus. Schrödinger solved his equation by doing what mathematical physicists always do in trying to solve differential equations of this kind, they apply boundary conditions.

Consider a guitar string that is bound at both ends, called the nut and the bridge respectively. Now strike the string to play an open string. As Figure 33 shows, the string will vibrate in an up and down motion in the form of half of one complete wave. But how else can the string vibrate? The answer is that it can also vibrate as two half wavelengths (a whole wavelength). This can be achieved on the guitar string by lightly placing a finger on the string, halfway along the fret-board (12th fret) and releasing it a fraction of a second after the string has been plucked in the region of the sound-hole by the other hand. The sound produced, if this is done properly, is a rather pleasant sounding bell-like tone which musicians call a harmonic. Additional modes in which the string can vibrate consist of 3, 4, 5, etc. half waves. The string may only vibrate as a whole number of half-waves. For example, there is no possible mode consisting of two and a half or three and one third half waves.

In mathematical terms, as well as physical terms, what has happened is that imposing boundary conditions (the two fixed points in the case of the actual guitar), has resulted in a set of motions that are characterized by whole numbers. This amounts to nothing less than quantization, or the restriction to whole number multiples of some value. By analogy, when Schrödinger applied such mathematical boundary conditions on his equation, the energies that he calculated for the electron turned out to be quantized, something that Bohr had been forced to smuggle into his equations from the outset. In Schrödinger's case, there is a substantial improvement in that he succeeded in deriving quantization of energy rather than merely introducing it artificially. This is the mark of a deeper treatment in which quantization emerges as a natural aspect of the theory.

Then there was another important ingredient that distinguished the new quantum mechanics from Bohr's old quantum theory. This second ingredient was discovered by the German physicist, Werner Heisenberg, who found the following very simple relationship for the uncertainty in position of a particle multiplied by the uncertainty for its momentum:

$$\Delta s . \Delta p = h / 4\pi$$

This equation implies that we must abandon the commonsense notion that a particle like an electron has a simultaneously well-defined position and momentum. What Heisenberg's relation holds is that, the more we pin-point the position of the electron, the less we are able to pin-point its momentum and vice versa. It is as though the motion of particles takes on a fuzzy or uncertain nature. Whereas Bohr's model had dealt with precise planetary-like orbits of electrons, circling the nucleus, the new view in quantum mechanics is that we may no longer speak of definite orbits for electrons. Instead, the theory retreats to talking in terms of probabilities (uncertainties) instead of certainties. When this view is combined with the idea that electrons are waves the picture that emerges is radically different from Bohr's model.

In quantum mechanics, the electron is considered to be spread throughout a spherical shell. It is as though the familiar particle has been converted into a gas which spreads to fill the entire contents of a sphere which corresponds roughly to a three-dimensional version of Bohr's two-dimensional orbit. In addition, the quantum mechanical sphere of charge does not have a sharp edge to it because of Heisenberg uncertainty. The first, or 1s orbital, which we have been discussing is in fact a volume that is associated with a 90% probability what we normally regard as a localized particle called the electron. And this is only the first of many solutions to the Schrödinger wave equation. Larger orbitals, as we move further from the nucleus, begin to have shapes other than spherical shells (Figure 34).

Now to consider the broader picture concerning quantum theory, quantum mechanics, and the periodic table. As we saw in Chapter 7, Bohr's original introduction of the concept of quantization into the study of atomic structure was carried out for the hydrogen atom. But in the same set of original papers, he began to explain the form of the periodic table by assuming that energy quantization also took place in many-electron atoms. The subsequent introduction of three further quantum numbers, in

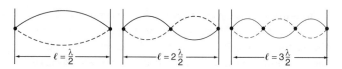

$\ell = \frac{\lambda}{2}$ $\ell = 2\frac{\lambda}{2}$ $\ell = 3\frac{\lambda}{2}$

33. Boundary conditions and the emergence of quantization

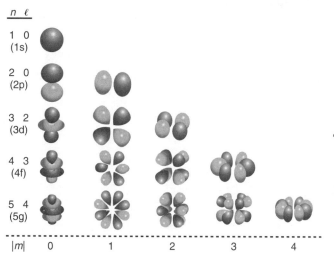

n	ℓ					
1	0 (1s)					
2	0 (2p)					
3	2 (3d)					
4	3 (4f)					
5	4 (5g)					

| $|m|$ | 0 | 1 | 2 | 3 | 4 |

34. s, p, d, and f orbitals

addition to Bohr's first quantum number, was also carried out in order to better explain the periodic table. And even before Bohr's venture into quantization concepts, the leading atomic physicist of the day, J. J. Thomson, had begun speculating about the arrangements of electrons in the various atoms that make up the periodic table.

The periodic table has therefore served as a testing ground for the theories of atomic physics and for many early aspects of quantum theory and the later quantum mechanics. The situation that exists today is that chemistry, and in particular the periodic table, is regarded as being fully explained by quantum mechanics. Even though this is not quite the case, the explanatory role that the theory continues to play is quite undeniable. But what seems to be forgotten in the current reductionist climate is that the periodic table led to the development of many aspects of modern quantum mechanics, and so it is rather short-sighted to insist that only the latter explains the former.

Chapter 9

Modern alchemy: from missing elements to synthetic elements

The periodic table consists of about 90 elements which occur naturally, ending with element 92, uranium. I say about 90 because one or two elements like technetium were first created artificially and only later found to occur naturally on earth.

Chemists and physicists have succeeded in synthesizing some of the elements that were missing between hydrogen (1) and uranium (92). In addition, they have synthesized a further 25 or so new elements beyond uranium, although, again, one or two of these, like neptunium and plutonium, were later found to exist naturally in exceedingly small amounts.

At the time of writing, the heaviest element for which there is good experimental evidence is element 118. All other elements between 92 and 118 have also been successfully synthesized, including element 117 which was announced in April of 2010.

The synthesis of many elements involved starting with a particular nucleus and subjecting it to bombardment with small particles with the aim of increasing the atomic number and hence changing the identity of the nucleus in question. More recently, the method of synthesis has changed to involve the collision of nuclei of considerable weights but still with the aim of forming a larger and heavier nucleus.

There is a fundamental sense in which all these syntheses are descended from a crucial experiment, conducted by Rutherford and Soddy, in 1919, at the University of Manchester. What Rutherford and colleagues did was to bombard nuclei of nitrogen with alpha particles (helium ions) with the result that the nitrogen nucleus was transformed into that of another element. Although they did not realize it initially, the reaction had produced an isotope of oxygen. Rutherford had achieved the first ever transmutation of one element into a completely different one. The dream of the ancient alchemists had become a reality, and this general process has continued to yield new elements right up to the present time.

$$^{14}_{7}N + ^{4}_{2}He \rightarrow ^{17}_{8}O + ^{1}_{1}H$$

However, this reaction did not produce a new element but just an unusual isotope of an existing element. Rutherford used alpha particles produced by the radioactive decay of other unstable nuclei such as uranium. It soon emerged that similar transmutations could be carried out with target atoms other than nitrogen but extending only as far as calcium, which has the atomic number of 20. If heavier nuclei were to be transmuted, it would require more energetic projectiles than naturally produced alpha particles.

The situation changed in the 1930s following the invention of the cyclotron by Ernest Lawrence at the University of California, Berkeley. This machine made it possible to accelerate alpha particles to hundreds and even thousands of times the speed possessed by naturally produced alpha particles. In addition, another projectile particle, the neutron, was discovered in 1932, having the added advantage of possessing a zero electric charge, which meant that it could penetrate a target atom without suffering any repulsion from the charged protons inside the nucleus.

Missing elements

In the mid-1930s, four gaps remained to be filled in the then existing periodic table. They consisted of the elements with atomic numbers of 43, 61, 85, and 87. Interestingly, the existence of three of these elements had been clearly predicted by Mendeleev many years before and called by him eka-manganese (43), eka-iodine (85), and eka-caesium (87). Three of these four missing elements were first discovered as a result of their being artificially synthesized in the 20th century.

In 1937, experiments were conducted at the Berkeley cyclotron facility in which a molybdenum target was bombarded with deuterium projectiles (isotopes of hydrogen with twice the mass of the more abundant isotope). One of the researchers was Emilio Segrè, a postdoctoral fellow from Sicily who then took the irradiated plates back to his native country. In Palermo, Segrè and Perrier analysed the plates and were able to confirm the formation of a new element, number 43, which they subsequently named technetium.

$$_{42}Mo + {}_{1}H \rightarrow {}_{43}Tc + {}_{0}n$$

This was the first completely new element obtained by transmutation, 18 years after Rutherford's classic experiment had demonstrated the possibility. Traces of the new element, technetium, were later found to occur naturally in the earth in vanishingly small quantities.

Element 85, or Mendeleev's eka-iodine, was the second missing element discovered as a result of artificial synthesis. Given its atomic number of 85, it could have been formed either from polonium (84) or bismuth (83). Polonium is very unstable and radioactive, and so attention turned to bismuth, which is actually the last stable element, while all those following it decay via radioactivity. And given that the atomic number of bismuth is two short of that of element 85, the obvious bombarding projectile would once again be an alpha particle. In 1940, Corson, Mackenzie, and Segrè, who had by then permanently settled in the USA, conducted just such an experiment in Berkeley and obtained an isotope of element 85 with a half-life of 8.3 hours. They named it astatine from the Greek *astatos*, meaning 'unstable'. The third of the missing elements to be artificially synthesized was element 61, also at the Berkeley cyclotron, this time the team consisting of Marinsky, Glendenin, and Coryell. The reaction involved collision of neodymium atoms with deuterium atoms.

Just to complete the story, on the fourth missing element, this had already been discovered by the French chemist Marguerite Perey, in 1939. She had started working as a laboratory assistant to Madame Curie and did not even have an undergraduate degree at the time of her discovery. She named her newly discovered element francium after her native country. This element did not require artificial synthesis but was found to be the by-product of the natural radioactive disintegration of actinium. Perey eventually rose to full professor and the directorship of the main institution for nuclear chemistry in France.

Transuranium elements

We finally come to the synthesis of the elements beyond the original 1 to 92. In 1934, three years before the synthesis of technetium, Enrico Fermi, working in Rome, began bombarding element targets with neutrons in the hope of synthesizing transuranium elements. Fermi believed that he had succeeded in producing two such elements, which he promptly named ausonium (93) and hesperium (94). But it was not to be.

After announcing the findings in his Nobel Prize acceptance speech of the same year, he quickly retracted the claim when it came to producing a written version of his presentation. The explanation for the erroneous claims emerged one year later, in 1938, when Otto Hahn, Fritz Strassmann, and Lise Meitner discovered nuclear fission. It became clear that, on collision with a neutron, the uranium nucleus, for example, could break up to form two middle-sized nuclei rather than a larger one.

For example, uranium was capable of forming caesium and rubidium by the following fission reaction:

$$_{92}U + {_0}n \rightarrow {_{55}}Cs + {_{37}}Rb$$

Fermi and his collaborators had been observing such products of nuclear fission processes instead of forming heavier nuclei as they first believed.

Real transuranium elements

The true identification of element 93 was finally carried out, in 1939, by Edwin McMillan and collaborators at Berkeley. It was named neptunium because it followed uranium in the periodic table just as the planet Neptune follows Uranus in terms of its distance from the Sun. A chemist in the team, Philip Abelson, discovered that element 93 did not behave as eka-rhenium, as expected from its presumed position in the periodic table. It was on the basis of this, and similar findings on element 94, or plutonium, that Glen Seaborg was to propose a major modification to the periodic table (See Chapter 1). As a result, the elements from actinium (89) onwards were no longer regarded as transition metals but as analogues of the lanthanide series. Consequently, there was no need for elements like 93 and 94 to behave like eka-rhenium and eka-osmium, since they had migrated to different places on the revised periodic table.

The synthesis of elements 94 to 97 inclusive, named plutonium, americium, curium, and berkelium, took place in the remaining years of the 1940s, and number 98, or californium, arrived in 1950. But this sequence looked as though it was about to end, since the heavier the nucleus, the more unstable it is, generally speaking. The problem became one of needing to accumulate enough target material in the hope of bombarding it with neutrons to transform the element into a heavier one. At this point, serendipity intervened. In 1952, a thermonuclear test explosion, code name Mike, was carried out close to the Marshall Islands in the Pacific Ocean. One of the outcomes was that intense streams of neutrons were produced, thus enabling such reactions as would not have been possible otherwise at this time. For example, the U-238 isotope can collide with 15 neutrons to form U-253, which subsequently undergoes the loss of seven beta particles, resulting in the formation of element 99, named einsteinium.

$$^{238}_{92}U + 15\,^{1}_{0}n \rightarrow\, ^{253}_{92}U \rightarrow\, ^{253}_{99}Es + 7\,_{-1}\beta$$

Element 100, named fermium, was produced in a similar manner as a result of the high-neutron flux produced by the same explosion and as revealed by analysis of the soil from the nearby Pacific islands.

From 101 to 106

Advancing further along the sequence of ever heavier nuclei required a quite different approach given that beta decay does not take place for elements above Z = 100. Many technological innovations were required, including the use of linear accelerators rather than cyclotrons, the former allowing researchers to accelerate highly intense beams of ions at well-defined energies. The projectile particles could also now be heavier than neutrons or alpha particles. During the Cold War, the only countries that possessed such facilities were the two superpowers of the USA and the Soviet Union.

In 1955, mendelevium, element 101, was produced in this way at the linear accelerator at Berkeley:

$${}^{4}_{2}He + {}^{253}_{99}Es \rightarrow {}^{256}_{101}Md + {}^{1}_{0}n$$

The possible combinations of nuclei became more numerous. For example, element 104, or rutherfordium, was made in Berkeley through the following reaction:

$${}^{12}_{6}C + {}^{249}_{98}Cf \rightarrow {}^{257}_{104}Rf + 4\,{}^{1}_{0}n$$

In Dubna, Russia, a different isotope of the same element was created in the reaction:

$${}^{22}_{10}Ne + {}^{242}_{94}Pu \rightarrow {}^{259}_{104}Rf + 5\,{}^{1}_{0}n$$

In all, six elements, 101 to 106, were synthesized by this approach. Partly as a result of Cold War tension between the USA and the Soviet Union, claims for the synthesis of most of these elements were hotly disputed, and these disputes continued for many years. But having reached element 106, a new problem arose that necessitated yet another new approach. It was at this point that German scientists entered the field, with the establishment of the GSI, Institute for Heavy-Ion Research in Darmstadt. The new technology was named 'cold-fusion' but had nothing to do with the kind of cold-fusion in a test tube as announced by the chemists Martin Fleischmann and Stanley Pons in 1989.

Cold-fusion in the transuranium field is a technique whereby nuclei are made to collide with one another at slower speeds than were previously used. As a result, less energy is generated, and consequently there is a decreased possibility that the combined nucleus can fall apart. This technique was originally devised by Yuri Oganessian, a Soviet physicist, but was developed more fully in Germany.

Elements 107 onwards

In the early 1980s, elements 107 (bohrium), 108 (hassium), and 109 (meitnerium) were successfully synthesized in Germany using the cold-fusion method. But then another roadblock became apparent. Many new ideas and techniques were tried while collaboration between the US, Germany, and what by now had become Russia, began to develop after the fall of the Berlin Wall and the Soviet Union. In 1994, after ten years of stagnation, the German lab in Darmstadt announced the synthesis of element 110 formed by the collision of lead and nickel ions:

$$^{208}_{82}Pb + ^{62}_{28}Ni \rightarrow ^{270}_{110}Ds \rightarrow ^{269}_{110}Ds + ^{1}_{0}n$$

The half-life of the resulting isotope was a mere 170 microseconds. The Germans called their element darmstadtium, not surprisingly given the previous naming of berkelium and dubnium by American and Russian teams respectively. A month later, the Germans had obtained element 111, which later became known as roentgenium after Röntgen, the discoverer of X-rays. February 1996 saw the synthesis of the next element in the sequence, 112, which was officially named copernicium in the year 2010 and remains the heaviest officially named element at the time of writing this book.

113 to 118

Since 1997, several claims have been published for the synthesis of elements 113 all the way to 118, the latest being element 117, synthesized in 2010. This situation is well understood given that nuclei with an odd number of protons are invariably more unstable than those with an even number of protons. This difference in stability occurs because protons, like electrons, have a spin of one half and enter into energy orbitals, two by two, with opposite spins. It follows that even numbers of protons frequently produce total spins of zero and hence more stable nuclei than those with unpaired proton spins as occurs in nuclei with odd numbers of protons such as 115 or 117.

The synthesis of element 114 was much anticipated because it had been predicted for some time that it would represent an 'island of stability', that is to say a portion of the table of nuclei with enhanced stability. Element 114 was first claimed by the

Dubna lab in late 1998, but only definitely produced in further experiments in 1999 involving the collision of a plutonium target with calcium-48 ions. The labs at Berkeley and Darmstadt have recently confirmed this finding. At the time of writing, something like 80 decays involving element 114 have been reported, 30 of which come from the decay of heavier nuclei such as 116 and 118. The longest-lived isotope of element 114 has a mass of 289 and a half-life of about 2.6 seconds, in agreement with predictions that this element would show enhanced stability.

On 30 December 1998, the Dubna-Livermore labs published a joint paper, claiming element 118 had been observed as a result of the following reaction:

$$_{36}Kr + {}_{82}Pb \rightarrow {}_{118}Uuo + {}_{0}n$$

After several failed attempts to reproduce this result in Japan, France, and Germany, the claim was officially retracted in July 2001. Much controversy followed, including the dismissal of one senior member of the research team who had published the original claim.

A couple of years later, new claims were announced from Dubna and were followed in 2006 by further claims by the Lawrence-Livermore Laboratory in California. Collectively, the US and Russian scientists made a stronger claim, that they had detected four more decays of element 118 from the following reaction:

$$_{98}Cf + {}_{20}Ca \rightarrow {}_{118}Uuo + 3\,{}_{0}n$$

The researchers are highly confident that the results are reliable, since the chance that the detections were random events was estimated to be less than one part in 100,000. Needless to say, no chemical experiments have yet been conducted on this element in view of the paucity of atoms produced and their very short lifetimes of less than one millisecond.

In 2010, an even more unstable element, number 117, was synthesized and characterized by a large team of researchers working in Dubna, as well as several labs in the USA. The periodic table has reached an interesting point at which all 118 elements either exist in nature or have been created artificially in special experiments. This includes a remarkable 26 elements beyond the element uranium, which has been traditionally regarded as the last of the naturally occurring elements. At the time of writing, there are even plans to attempt the creation of yet heavier elements such as 119 and 120, and there are no reasons for thinking that there should be any immediate end to the sequence of elements that can be formed.

Chemistry of the synthetic elements

The existence of superheavy elements raises an interesting new question and also a challenge to the periodic table. It also affords an intriguing new meeting point for theoretical predictions to be pitted against experimental findings. Theoretical calculations suggest that the effects of relativity become increasingly important as the nuclear charge of atoms increases. For example, the characteristic colour of gold, with a rather modest atomic number of 79, is now explained by appeal to relativity theory. The larger the nuclear charge, the faster the motion of inner shell electrons. As a consequence of gaining relativistic speeds, such inner electrons are drawn closer to the nucleus, and this in turn has the effect of causing greater screening on the outermost electrons which determine the chemical properties of any particular element. It has been predicted that some atoms should behave chemically in a manner that is unexpected from their presumed positions in the periodic table.

Relativistic effects thus pose the latest challenge to test the universality of the periodic table. Such theoretical predictions have been published by various researchers over a period of many years, but it was only when elements 104 and 105, rutherfordium and dubnium respectively, were chemically examined that the

	3	4	5	6	7	8	9	10	11	12
	Sc	Ti	V	Cr	Mn	Fe	Co	Ni	Cu	Zn
	Y	Zr	Nb	Mo	Tc	Ru	Rh	Pd	Ag	Cd
	Lu	Hf	Ta	W	Re	Os	Ir	Pt	Au	Hg
	Lr	Rf	Db	Sg	Bh	Hs	Mt	Ds	Rg	Cn

35. Fragment of periodic table showing groups 3 to 12 inclusive

situation reached something of a climax. It was found that the chemical behaviour of rutherfordium and dubnium was in fact rather different from what one would expect intuitively from where these elements lie in the periodic table. Rutherfordium and dubnium did not seem to behave like hafnium and tantalum, respectively, as they should have done.

For example, in 1990, K. R. Czerwinski, working at Berkeley, reported that the solution chemistry of element 104, or rutherfordium, differed from that of zirconium and hafnium, the two elements lying above it. Meanwhile, he also reported that rutherfordium's chemistry resembled that of the element plutonium which lies quite far away in the periodic table. As to dubnium, early studies showed that it too was not behaving like the element above it, namely tantalum (Figure 35). Instead, it showed greater similarities with the actinide protactinium. In other experiments, rutherfordium and dubnium seemed to be behaving more like the two elements above hafnium and tantalum, namely zirconium and niobium.

It was only when the chemistry of the following elements, seaborgium (106) and bohrium (107), was examined that they showed that the expected periodic behaviour was resumed. The titles of the articles that announced these discoveries spoke for themselves. They included, 'Oddly Ordinary Seaborgium' and 'Boring Bohrium', both references to the fact that it was business as usual for the periodic table. Even though relativistic effects should be even more pronounced for these two elements, the expected chemical behaviour seems to outweigh any such tendencies.

The fact that bohrium behaves as a good member of group 7 can be seen from the following argument that I have proposed. This approach also represents a kind of 'full circle' since it involves a triad of elements. As the reader may recall from Chapter 3, the discovery of triads was the very first hint of a numerical regularity relating the properties of elements within a common group. Here is the data for measurements carried out on the standard sublimation enthalpies, of analogous compounds of technetium, rhenium, and bohrium with oxygen and chlorine (energy required to convert a solid directly into a gas).

TcO_3Cl = 49 kJ/mol
ReO_3Cl = 66 kJ/mol
BhO_3Cl = 89 kJ/mol

36. Sublimation enthalpies for elements in group 7 showing that element 107 is a genuine member of this group

Predicting the value for BhO_3Cl using the triad method gives 83kJ/mol, or an error of only 6.7% compared with the above experimental value of 89kJ/mol. This fact lends further support to the notion that bohrium acts as a genuine group 7 element.

Group 7

The challenge to the periodic law from relativistic effects became even more poignant in the case of number 112, or copernicium, the most recent element for which chemical experiments have been conducted. Once again, relativistic calculations indicated modified chemical behaviour to the extent that the element was thought to behave like a noble gas rather than like mercury below which it is placed in the periodic table. Experiments carried out on sublimation enthalpies on element 112 then showed that, contrary to earlier expectations, the element truly belongs in group 12 along with zinc, cadmium, and mercury.

Element 114 presented a similar story with early calculations and experiments suggesting noble gas behaviour but more recent experiments supporting the notion that the element behaves like the metal lead as expected from its position in group 14. The conclusion would seem to be that chemical periodicity is a remarkably robust phenomenon. Not even the powerful relativistic effects due to fast-moving electrons seem to be capable of toppling a simple scientific discovery that was made around 150 years ago.

Chapter 10
Forms of the periodic table

Much has been said about the periodic table in previous chapters, but one important aspect has not yet been addressed. The question is why have so many periodic tables been published and why are there so many currently on offer in textbooks, articles, and on the Internet? Is there such a thing as an 'optimal' periodic table? Does such a question even make sense, and if so, what progress has been made towards identifying such an optimal table?

In his classic book on the history of the periodic table, Edward Mazurs included illustrations as well as references to about 700 periodic tables that have been published since the periodic table was first assembled in the 1860s. In the 40 or so years that have elapsed since the publication of Mazurs' book, at least another 300 tables have appeared, especially if we include new periodic systems posted on the Internet. The very fact that so many periodic tables exist is something that demands an explanation. Of course, many of these tables may not have anything new to offer, and some are even inconsistent from a scientific point of view. But even if we were to eliminate these misleading tables, a very large number of tables still remain.

In Chapter 1, we saw that there are three basic forms of the periodic table, the short form, the medium-long form, and the long

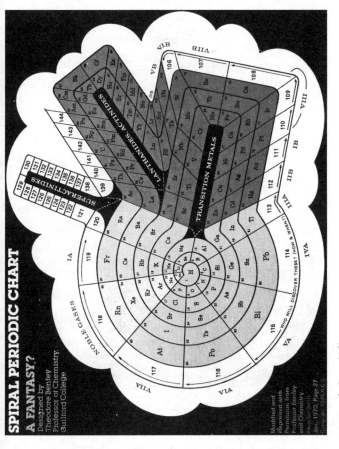

37. Benfey's periodic system

38. Dufour's periodic tree

form. All three of these convey very much the same information, although, as was noted, the grouping of elements with the same valence is treated differently in each of these formats. In addition, there are periodic tables that don't look like tables in the literal sense of having a rectangular shape. One important variation of this kind includes the circular and elliptical periodic systems which serve to highlight the continuity of the elements in a way that is perhaps better than the rectangular forms do. In circular or elliptical systems, there are no discontinuities at the end of the periods as there are in the rectangular formats between, say, neon and sodium or argon and potassium. But the lengths of periods vary, unlike the periods on a clock face. As a result, the circular periodic table designer needs to accommodate the longer periods which have transition elements in some way. For example, Benfey's table (Figure 37) does this by using bulges for the transition metals which jut off the main circular system. Then there are three-dimensional periodic tables such as the one designed by Fernando Dufour from Montreal in Canada (Figure 38).

But I am going to suggest that all these variations only involve changing the shape of the periodic system and that there is no fundamental difference between them. What does constitute a major variation is the placement of one or more elements in a different group than where it is usually located in the conventional table. But before taking up this point, let me pause to discuss the design of the periodic table in general.

The periodic table is a deceptively simple concept, or at least that's how it appears. This is what invites amateur scientists to try their hand at developing new versions, which they often claim have properties that are superior to all previously published systems. Moreover, there have been a number of cases in which amateurs, or outsiders to the fields of chemistry and physics, have indeed made major contributions. For example, Anton van den Broek, who was mentioned in Chapter 6, was an economist who first came up with the notion of atomic number, an idea that he developed in several articles in such journals as *Nature* magazine. A second example was Charles Janet, a French engineer, who in 1929 published the first known version of the left-step periodic system that continues to command much attention among periodic table amateurs as well as experts (Figure 39).

																														H	He	1
																														Li	Be	2
																								B	C	N	O	F	Ne	Na	Mg	3
																								Al	Si	P	S	Cl	Ar	K	Ca	4
														Sc	Ti	V	Cr	Mn	Fe	Co	Ni	Cu	Zn	Ga	Ge	As	Se	Br	Kr	Rb	Sr	5
														Y	Zr	Nb	Mo	Tc	Ru	Rh	Pd	Ag	Cd	In	Sn	Sb	Te	I	Xe	Cs	Ba	6
La	Ce	Pr	Nd	Sm	Eu	Gd	Tb	Dy	Ho	Er	Tm	Yb	Lu	Hf	Ta	W	Re	Os	Ir	Pt	Au	Hg	Tl	Pb	Bi	Po	At	Rn	Fr	Ra		7
Ac	Th	Pa	U	Pu	Am	Cm	Bk	Cf	Es	Fm	Md	No	Lr	Rf	Db	Sg	Bh	Hs	Mt	Ds	Rg	Cn										8

39. The left-step periodic table by Charles Janet

Now, what about the other question that was raised earlier? Does it even make sense to be seeking an optimal periodic table, or are the amateurs deluding themselves along with the experts who devote time to this question? The answer to this question lies, I believe, in one's philosophical attitude towards the periodic system. If one believes that the approximate repetition in the properties of the elements is an objective fact about the natural world, then one is adopting the attitude of a realist. To such a person, the question of seeking an optimal periodic table does make perfect sense. The optimal periodic table is the one which best represents the facts of the matter concerning chemical periodicity, even if this optimal table may not yet have been discovered.

On the other hand, an instrumentalist or an anti-realist about the periodic table might believe that the periodicity of the elements is a property that is imposed on nature by human agents. If this is the case, there is no longer a burning desire to discover *the* optimal table since no such thing would even exist. For such a conventionalist or anti-realist, it does not matter precisely how the elements are represented since he or she believes that one is dealing with an artificial rather than a natural relationship among the elements.

Just to declare my own position, I am very much a realist when it comes to the periodic table. For example, it surprises me that many chemists take an anti-realistic stance concerning the periodic table; when asked whether the element hydrogen belongs to group 1 (alkali metals) or group 17 (the halogens), some chemists respond that it does not matter.

There are some final general issues to mention before we examine the details of alternative tables and possible optimal tables. One is the question of utility of this or that periodic table. Many scientists tend to favour one or other particular form of the periodic table because it may be of greater use to them in their scientific work as an astronomer, a geologist, a physicist, and so on. These are the tables that are primarily based on utility. There are other tables that seek to highlight the 'truth' about the elements, for want of a better expression, rather than usefulness to some kinds of scientists. Needless to say, any quest for an optimal periodic table should avoid the question of utility, especially if it is utility to one particular scientific discipline or subdiscipline. Moreover, a table that strives for the truth about the elements, above all, will hopefully turn out to be useful to various disciplines if it succeeds in somehow capturing the true nature and relationship between the elements. But such usefulness would come as a bonus as it is not something that should determine the manner in which the optimal table is arrived at.

There is also the important question of symmetry. This too is a rather tricky issue. Many proponents of alternative periodic tables claim that their table is superior because the elements are represented in a more symmetrical, regular, or somehow more elegant or beautiful fashion. The question of symmetry and beauty in science has been much debated, but as with all aesthetic questions, what may appear beautiful to one person may not be regarded in the same way by somebody else. Also, one must beware of imposing beauty or regularity on nature where it might not actually be present. Too many proponents of alternative tables seem to argue exclusively about the regularity of their representation and sometimes forget that they are talking about the representation and not the chemical world itself.

Some specific cases

Having mentioned all these preliminaries, we can dive into some proposed new tables, assuming of course that, as I am suggesting, it does make sense to seek an optimal periodic table. Let us begin with the left-step periodic table which is one of those substantially different periodic systems in which elements are placed in different groups than in the more conventional tables. The left-step table was first proposed by Charles Janet, in 1929, shortly after the development of quantum mechanics. However, it seems that Janet's proposal owes nothing to quantum mechanics but was based entirely on aesthetic grounds. But it soon became clear that there are some major features of the left-step table which may correlate rather better with the quantum mechanical account of atoms than the conventional tables do.

First of all, what exactly is the left-step table and how does it differ from other periodic tables? The left-step table is obtained by moving the element helium from the top of the noble gases (group 18) to the top of the alkaline earth metals (group 2). The entire two groups on the left of this table are then moved to the right-hand edge to form a new table. In addition, the block of 28 elements, the rare earths, that are usually presented as a form of footnote to the periodic table, are moved to the left side of the new table. As a result of this move, the rare earth elements become fully incorporated into the periodic table to the left of the transition metal block.

Among the virtues of the new table are the fact that the overall shape is more regular and more unified. In addition, we now obtain two very short periods rather than just one as seen in the usual periodic tables. So rather than having just one anomalous period length that does not repeat, the left-step table features all period lengths repeating once to give a sequence of 2, 2, 8, 8, 18, 18, etc. None of these advantages involve quantum mechanics but are features that Janet, who was not aware of this theory, appreciated. As we have seen in Chapter 8, the introduction of quantum mechanics to the periodic table resulted in an understanding based upon electronic configurations. In this approach, the elements in the periodic table differ from each other according to the type of orbital occupied by the differentiating electron (the last electron to enter the atom in the building-up process).

In the conventional tables, the elements in the two left-most groups are said to form the s-block because their differentiating electrons enter an s-orbital. Moving towards the right, we encounter the d-block, then the p-block, and finally the f-block, the last of which lurks underneath the main body of the table. This order of blocks from left to right is not the most 'natural' or expected one, since in every shell the distance from the nucleus follows the order,

$$s < p < d < f$$

The left-step table maintains this ordering, although it is in reverse order. Whether this is really an advantage is debatable, however, since the order of filling of orbitals takes the form

$$s < f < d < p$$

which is perfectly in keeping, at least with the traditional long-form table, as the blocks are arranged from left to right. Moreover, it is the order of filling rather than the distance of electrons in various types of orbitals from the nucleus that should be regarded as being more fundamental.

But there may be another advantage from the quantum mechanical point of view. There is no disputing the fact that the electronic configuration of the helium atom shows two electrons, both of which are in a 1s orbital. This should make helium an s-block element. But in conventional periodic tables, helium is placed among the noble gases because of its chemical properties, being highly inert just like the remaining noble gases (Ne, Ar, Kr, Xe, Rn).

This situation seems to provide a parallel with the earlier discussed historical case of the tellurium–iodine reversal in which atomic weight ordering had to be ignored in order to preserve chemical similarities. Similarly, in the case of helium, there appear to be two possibilities:

1. Electronic structure is not the final arbiter over the placement of elements into groups and may well be replaced by some new criterion in due course (for example, atomic weight was eventually replaced by atomic number in the ordering of the elements, thus solving the pair reversal problem).

2. We do not really have a parallel case and electronic configuration still rules the day, in which case the apparently inert chemistry of helium should be ignored.

Notice that option 1 actually favours the conventional periodic tables, whereas option 2 favours the left-step table. Clearly, it is not so easy to decide whether the left-step table presents an advantage from the quantum mechanical point of view. Let me now throw another idea into the mix. Recall what was said in Chapter 4 on the nature of the elements and how Mendeleev, in particular, favoured regarding elements in the more abstract sense, rather than being tied to the elements as being simple or isolated substances. This reliance on the abstract sense of an element could be used to justify moving helium to the alkaline earth group. The concern that helium's chemical unreactivity denies it a place among the more reactive alkaline earths can thus be countered by putting one's attention on the nature of the element as an abstract entity rather than on its chemical properties. However, this move amounts to saying, 'why not' place helium in the alkaline earths if its chemistry can be ignored?

A new criterion: preserving or creating new triads

There may be a stronger and more positive criterion ('why yes' rather than 'why not') for deciding on the placement of helium, or any other element for that matter. This new criterion represents something of a full-circle as far as the development of the periodic table is concerned. Recall how the first hint of any numerical regularity among the elements came in 1817 with Döbereiner's discovery of triads of elements such as lithium, sodium, and potassium where the atomic weight of the middle element has a value intermediate between the two flanking elements. But the ordering principle consisting of atomic weight was replaced by that of atomic number. What does this change do for the existence of triads of elements in the periodic table? Well, if anything, it strengthens the concept, since the former approximate atomic weight triads now become exact atomic number triads. Consider the following two classic atomic weight triads, using atomic weights cited to 4 or 5 significant figures.

Li	6.940	Cl	35.45
Na	22.99 (cf. average of 23.02)	Br	79.90 (cf. average of 81.18)
K	39.10	I	126.90

And now consider the corresponding atomic number triads,

Li	3	Cl	17
Na	11	Br	35
K	19	I	53

Whereas the atomic weights of sodium and bromine correspond only approximately with the averages obtained from the weights of their flanking elements, the atomic numbers of sodium and bromine are the exact averages of those of the flanking elements.

What if we were to try to settle the placement of the element helium by considering atomic number triads? The outcome of applying this approach is very clear and obvious. If helium is left in its traditional place among the noble gases, it already forms part of a perfect atomic number triad.

He	2
Ne	10
Ar	18

If helium is moved to the alkaline earth group, however, as is done in the left-step table, we only succeed in destroying a perfectly good atomic number triad.

He	2
Be	4
Mg	12

Same criterion applied to other elements that are difficult to place

Another element which has traditionally caused problems is the very first one, namely hydrogen. In chemical terms, it seems to belong to group 1 (alkali metals) due to its ability to form +1 ions of H^+. Alternatively, H is rather unique in that it can also form an H^- ion, as in the case of metal hydrides such as NaH, CaH_2, etc. This behaviour supports the placement of hydrogen in group 17 (the halogens) which mainly also form -1 ions. How can this issue be settled categorically? Some authors take an easy way out by allowing hydrogen to 'float' majestically above the main body of the periodic table; in other words, they do not commit themselves to either one of the possible placements.

This has always seemed to me a display of 'chemical elitism' since it seems to imply that whereas all the elements are subject to the periodic law, hydrogen is somehow a special case and as such is above the law, very much like the British Royal family once was. What about trying to use the atomic number triads criterion in order to settle the question of the placement of hydrogen? As in the case of helium, the outcome of using this approach produces a very clear-cut result in favour of placing hydrogen among the halogens rather than the alkali metals. In the conventional periodic table, in which hydrogen is placed in the alkali metals, there is no perfect triad, whereas if hydrogen is allowed to sit at the top of the halogens, a new atomic number triad comes into being.

																												H	He	Li	Be
																								B	C	N	O	F	Ne	Na	Mg
																								Al	Si	P	S	Cl	Ar	K	Ca
														Sc	Ti	V	Cr	Mn	Fe	Co	Ni	Cu	Zn	Ga	Ge	As	Se	Br	Kr	Rb	Sr
														Y	Zr	Nb	Mo	Tc	Ru	Rh	Pd	Ag	Cd	In	Sn	Sb	Te	I	Xe	Cs	Ba
La	Ce	Pr	Nd	Pm	Sm	Eu	Gd	Tb	Dy	Ho	Er	Tm	Yb	Lu	Hf	Ta	W	Re	Os	Ir	Pt	Au	Hg	Tl	Pb	Bi	Po	At	Rn	Fr	Ra
Ac	Th	Pa	U	Np	Pu	Am	Cm	Bk	Cf	Es	Fm	Md	No	Lr	Rf	Db	Sg	Bh	Hs	Mt	Ds	Rg	Cn								

40. Periodic table based on maximizing atomic number triads

H	1		H	1	
Li	3	$(1 + 11)/2 \neq 3$	F	9	$(1 + 17)/2 = 9$
Na	11		Cl	17	

Now let me bring together the triad-driven findings on helium and hydrogen. If we regard atomic number triads as being fundamental, as I am recommending, then helium should remain in group 18, whereas hydrogen should be moved to group 17. The resulting periodic table that I have proposed in a number of articles takes the form shown in Figure 40.

Atomic number triads applied to group 3

There is also a long-standing dispute among chemists and chemical educators regarding group 3 of the periodic table. Some old periodic tables show the following elements in group 3:

Sc
Y
La
Ac

More recently, periodic tables shown in many textbooks began to show group 3 as:

Sc
Y
Lu
Lr

with arguments based on presumed electronic configurations. In 1986, an influential article was published by William Jensen of the University of Cincinnati, in which he argued, rather persuasively, that textbook authors and periodic table designers should return to showing group 3 as: Sc, Y, Lu, and Lr.

Then even more recently, a rearguard action occurred in which some authors have argued for a return to Sc, Y, La, Ac. How, if at all, does the notion of atomic number triads bear on this group 3 issue? If we consider atomic number triads, the answer is once again swift and categorical, namely that Jensen's grouping is favoured. Whereas the first triad below is exact,

```
Y    39
Lu   71 = (39 + 103)/2
Lr   103
```

the second triad is incorrect,

```
Y    39
La   57 does not equal (39 + 89)/2 = 64
Ac   89
```

But there is another reason why Jensen's arrangement is the better one, and this does not depend on any allegiance to atomic number triads.

If we consider the long-form periodic table and attempt to incorporate either lutetium and lawrencium or lanthanum and actinium into group 3, only the first placement makes any sense because it results in a continuously increasing sequence of atomic numbers. Conversely, the incorporation of lanthanum and actinium into group 3 of a long-form table results in two rather glaring anomalies in terms of sequences of increasing atomic numbers (Figure 41).

Finally, there is actually a third possibility, but this involves a rather awkward subdivision of the d-block elements as shown in Figure 42.

Although some books include a periodic table arranged like Figure 42, this is not a very popular design and for rather obvious reasons. Presenting the periodic table in this fashion requires that

Long-form periodic table with Lu and Lr in group 3:

1	2	3	4	5	6	7	8	9	10	11	12	13	14	15	16	17	18
1 H																	2 He
3 Li	4 Be											5 B	6 C	7 N	8 O	9 F	10 Ne
11 Na	12 Mg											13 Al	14 Si	15 P	16 S	17 Cl	18 Ar
19 K	20 Ca	21 Sc	22 Ti	23 V	24 Cr	25 Mn	26 Fe	27 Co	28 Ni	29 Cu	30 Zn	31 Ga	32 Ge	33 As	34 Se	35 Br	36 Kr
37 Rb	38 Sr	*39 Y*	40 Zr	41 Nb	42 Mo	43 Tc	44 Ru	45 Rh	46 Pd	47 Ag	48 Cd	49 In	50 Sn	51 Sb	52 Te	53 I	54 Xe
55 Cs	56 Ba	*71 Lu*	72 Hf	73 Ta	74 W	75 Re	76 Os	77 Ir	78 Pt	79 Au	80 Hg	81 Tl	82 Pb	83 Bi	84 Po	85 At	86 Rn
87 Fr	88 Ra	*103 Lr*	104 Rf	105 Db	106 Sg	107 Bh	108 Hs	109 Mt	110 Ds	111 Rg							

57 La	58 Ce	59 Pr	60 Nd	61 Pm	62 Sm	63 Eu	64 Gd	65 Tb	66 Dy	67 Ho	68 Er	69 Tm	70 Yb
89 Ac	90 Th	91 Pa	92 U	93 Np	94 Pu	95 Am	96 Cm	97 Bk	98 Cf	99 Es	100 Fm	101 Md	102 No

Long-form periodic table with Lu and Lr in group 3.

Long-form periodic table with La and Ac in group 3:

1	2	3	4	5	6	7	8	9	10	11	12	13	14	15	16	17	18
1 H																	2 He
3 Li	4 Be											5 B	6 C	7 N	8 O	9 F	10 Ne
11 Na	12 Mg											13 Al	14 Si	15 P	16 S	17 Cl	18 Ar
19 K	20 Ca	21 Sc	22 Ti	23 V	24 Cr	25 Mn	26 Fe	27 Co	28 Ni	29 Cu	30 Zn	31 Ga	32 Ge	33 As	34 Se	35 Br	36 Kr
37 Rb	38 Sr	*39 Y*	40 Zr	41 Nb	42 Mo	43 Tc	44 Ru	45 Rh	46 Pd	47 Ag	48 Cd	49 In	50 Sn	51 Sb	52 Te	53 I	54 Xe
55 Cs	56 Ba	*57 La*	72 Hf	73 Ta	74 W	75 Re	76 Os	77 Ir	78 Pt	79 Au	80 Hg	81 Tl	82 Pb	83 Bi	84 Po	85 At	86 Rn
87 Fr	88 Ra	*89 Ac*	104 Rf	105 Db	106 Sg	107 Bh	108 Hs	109 Mt	110 Ds	111 Rg							

58 Ce	59 Pr	60 Nd	61 Pm	62 Sm	63 Eu	64 Gd	65 Tb	66 Dy	67 Ho	68 Er	69 Tm	70 Yb	71 Lu
90 Th	91 Pa	92 U	93 Np	94 Pu	95 Am	96 Cm	97 Bk	98 Cf	99 Es	100 Fm	101 Md	102 No	103 Lr

Long-form periodic table with La and Ac in group 3.

41. Long-form periodic tables showing alternative placements of Lu and Lr. Only the upper version preserves a continuous sequence of atomic numbers

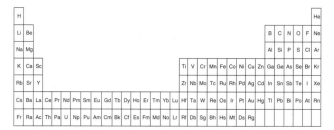

42. Third option for presenting long-form periodic table in which the d-block is split into two uneven portions of one and nine groups

the d-block of the periodic table should be split up into two highly unequal portions containing a block that is only one element wide and another block containing a nine-element-wide block. Given that this behaviour does not occur in any other block in the periodic table, it would appear to be the least likely of the three possible tables to reflect the actual arrangement of the elements in nature.

Returning to hydrogen and helium, let me make one final point. In terms of chemical and physical evidence I don't think we can yet settle the issue of the definitive periodic table or whether the left-step table is superior to a traid motivated table.

I hope that the reader comes away from this book with the realization that the periodic table is still very much a subject of interest and development, and might be encouraged to look at the suggested additional reading list.

Further reading

H. Aldersey-Williams, *Periodic Tales* (Viking, 2011).

P. Ball, *The Elements: A Very Short Introduction* (Oxford University Press, 2004).

M. Gordin, *A Well-Ordered Thing* (Basic Books, 2004).

T. Gray, *The Elements: A Visual Exploration of Every Known Atom in the Universe* (Black Dog and Leventhal, 2009).

S. Kean, *The Disappearing Spoon: And Other True Tales of Madness, Love, and the History of the World from the Periodic Table of the Elements* (Little, Brown and Company, 2010).

E. Mazurs, *Graphical Representations of the Periodic System during 100 Years* (University of Alabama Press, 1974).

E. R. Scerri, *The Periodic Table, Its Story and Its Significance* (Oxford University Press, 2007).

E. R. Scerri, *Selected Papers on the Periodic Table* (Imperial College Press, 2009).

J. W. van Spronsen, *The Periodic System of Chemical Elements: A History of the First Hundred Years* (Elsevier, 1969).

Recommended websites

Eric Scerri's website for history & philosophy of chemistry and the periodic table,

http://ericscerri.com/

Webelements, the leading periodic website, maintained by Mark Winter of Sheffield University,

http://www.webelements.com/

Mark Leach's metasynthesis site. A wonderful compendium of periodic tables,

http://www.meta-synthesis.com/webbook/35_pt/pt_database.php

Index

A

Abelson, Philip 113
abstract elements 31, 62, 131
acids
    and bases 20–1
    and metals 21
actinium (89) as start of rare earths 24
alchemy 2
    modern 110
Aldersey-Williams, Hugh 6
alpha particles 75
    in transmutation 110, 112
alphons 78
americium (95) 24
Ampère, André 35
argon (18)
    classification 69–71
    electrons 90
Aristotle 1
Armstrong, Henry 70
arsenic (33) and
        superconductivity 11
artificial elements 5
astatine (85), synthesis 111–12
atomic number 22, 77–80
    and isotopes 82–3
    triads 131–7
atomic weight 20–2, 32–3
    accuracy 36
    of argon 69–71

as fundamental ordering
        principle 64–5
    and Mendeleev 61
atoms, nuclear model 74–6
Avogadro, Amedeo 34–5

B

Barkla, Charles 77
bases and acids 20–1
Becquerel, Henri 73, 76–7
Benfey 123, 125
beryllium (4) and medicine 11
Berzelius, Jöns Jacob 9
bismuth (83) and
        transmutation 111–12
'black body radiation' 84–5
blocks on the periodic table 129–30
Bohr, Niels 27–8, 78, 85–8, 99, 101
bohrium (107) 115, 120
Boisbaudran, Paul Émile Lecoq de 45
bonding
    of atoms 94–5
    of elements 61–2
boron (5) and quantum theory 86
Brauner, Bohuslav 65
Broek, Anton van den 77–9, 125
bromine (35)
    as part of a triad 37
    properties 16
Bunsen, Robert 51

burning 30–1
Bury, Charles 97

C

cadmium (48) and medicine 11
caesium (55), name 7
calcium (20)
  configuration 91
  spectral lines 52
Cannizzaro, Stanislao 35, 42
carbon (6), properties 16
Chancourtois, Émile Béguyer
  de 43–5
chemical bonding 61–2
chemical periodicity 19–20
chlorine (17)
  name 7
  as part of a triad 37
  properties 16
  in sodium chloride 61–2
circular periodic tables 123, 125
classification, secondary vs.
  primary 24–6
cold-fusion method of synthesis 115
colours as element names 7
compounds for
  superconductivity 10–11
'conjugated triads' 40
conservation of matter 31
copernicium (112) 116, 121
copper (29), symbol 9
Corson 112
covalent bonding 94
Crookes, William 7, 47, 70
Curie, Marie 73–4, 77
curium (96) 24
cyclotron 110–11
Czerwinski, K. R. 119

D

d-block 130, 137
Dalton, John 9, 21, 31–3, 35
darmstadtium (110) 116

Davy, Humphry 5
Davy medal 68
deuterium projectiles 111, 112
diatomic molecules 34–5
divalent atoms 21
divisibility of atoms 34
Döbereiner, Wolfgang 36–7, 131
Dufour, Fernando 124, 125

E

Einstein, Albert 27, 85
einsteinium (99) 114
eka- elements 66, 68, 111
electrical energy and quantum
  theory 85
electricity to identify elements 5
electrons
  and chemical bonding 94–5
  configurations 97
  discovery 73, 92–3
  and quantum mechanics
    99–102, 106–8
  and quantum theory 27–8, 85–92
  relativistic effects 27
  as waves 104–5
'element', senses of the word 62–3
element *114* (unnamed) 116–17, 121
element *117* (unnamed) 117–18
element *118* (unnamed) 117
elements
  ancient Greek 1–3
  as basic substance 62, 63
  definition 3
  discovery 5
  names 6–9
  new 26
    isotopes as 80–1
    predictions 63, 65–7
  as simple substance xv, 3, 4, 62, 63
  synthesized 109–18
    behaviour 118–21
energy and quantum theory 85–6
enhanced stability 116–17

equivalent weight 20–2, 31–2
ether 1
existence of atoms 72

F

f-block 130
Fermi, Enrico 112–13
fermium (100) 114
fluorine (9), properties 16
forms of the periodic table 13–14
francium (87), discovery 112

G

gallium (31) 66
gases
  inert 69–71
  volumes 33–4
  see also halogens; noble gases
Gay-Lussac, Joseph-Louis 33–4
germanium (32) 66
  prediction of 55
  properties 16
Gmelin, Leopold 38–9
gold (79) and medicine 11
Greek philosophers 1–3
  notion of abstract elements 31
group *1* 15
group *2* 15
group *3*, triads 134–7
group *14*, properties 16
group *17*, properties 16–17, 19
group *18 see* noble gases
groups (columns) 12–13

H

Hahn, Otto, Fritz Strassmann, and
    Lise Meitner 113
halogens
  properties 19
  as a triad 37
Hartog, Philip 45
hassium (108) 115

Heisenberg, Werner 105–6
helium (2)
  in left-step periodic table 130–1
  name 6
  placement 132, 134
Hevesy, Georg von 81
Hinrichs, Gustavus 49–53
horizontal relationships 39–41
horizontal rows 12
Humboldt, Alexander von 33–4
Hund 99
hydrogen (1)
  as basic unit of atoms 35–6
  placement 127, 133–4
  and quantum theory 85
  reaction with oxygen 33–5
hydroxides in water 20

I

identification of elements 3
inert gases 69–71
International Union of Pure and Applied
    Chemistry (IUPAC) 8, 12
iodine (53)
  as part of a triad 37
  properties 16
  and tellurium 64–5, 70, 82–3
ionization energy 101–3
ions, positive vs. negative 20
isotopes 22, 80–3
  synthesis 110

J

Janet, Charles 125–6, 128–9
Jensen, William 134–5

K

Karlsruhe conference (1860) 42,
    59
Kean, Sam 6
Kirchoff, Gustav 51
Kremers, Peter 39–41

L

Langmuir, Irving 96
lanthanum (57) 24
Lapparent, Albert Auguste 45
Latin names 9
Lavoisier, Antoine 3–4, 30–1, 62
law
    of chemistry 19–20
    of conservation of matter 31
    looking for exceptions 26
    of multiple proportions 33
    of octaves 47–8
Lawrence, Ernest 110
lawrencium (103) as transition
        element 24
lead (82)
    isotopes 81–2
    properties 16
left-step periodic table 126,
        128–31
Lenssen, Ernst 37–8
Levi, Primo 6
Lewis, G. N. 92–5
light emissions 51–3
lithium (7)
    configuration 86
    as part of a triad 36–7
long-form periodic tables 14,
        17–18
Lothar Meyer, Julius 47, 53–7
lutetium (71) as transition
        element 24

M

Mackenzie 112
magnesium (12) and medicine 11
many-electron atoms 99–100, 106
Marinsky, Glendennin, and
        Coryell 112
Mazurs, Edward 122
McCoy, Herbert 81
McMillan, Edwin 113
medicine 11–12

medium-long-form periodic
        tables 14–17
meitnerium (109) 115
Mendeleev, Dimitri 3, 15, 40, 57,
        111
    classification of argon 71
    design of the periodic
        table 60–1
    life of 58–60
    predictions 63, 65–8
    reversal of tellurium and
        iodine 64–5
    senses of the term
        'element' 62–3
mendelevium (101) 114–15
mercury (80), symbol 8–9
metals
    and acids 21
    appearance 12
    vs. non-metals 20
    reactivity 15
Mike (thermonuclear test
        explosion) 114
Moseley, Henry 65, 73, 79–80
Mulliken 99
multiple proportions 33

N

Nagaoka, Hantaro 75
names of elements 6–9
neptunium (93) 113
neutron particles in
        transmutation 110–11
neutrons and atomic number 22
new elements 26
    isotopes as 80–1
    predictions 63, 65–8
Newlands, John 45–8
noble gases 17
    classification 69–71
    electrons 91–2
nuclear charge 78
nuclear fission 113
    to identify elements 5

nuclear model of the atom 74–6
nuclear reaction and transuranium
        elements 114–18

O

octaves, law of 47–8
Odling, William 48–9
Ogawa, Seiji 80
optimal periodic table 127–37
orbitals 28, 100, 102, 106–7
oxides in water 20
oxygen (8)
    atomic weight 32
    reaction with hydrogen 33–5
    synthesis 110
oxypnictides 11

P

p-block 130
pair reversals 64–5, 69–70, 71
    and isotopes 82–3
palladium (46), name 6
Paneth, Friedrich 81
Pauli, Wolfgang 88
Perey, Marguerite 112
periodic law 19–20
'periodic system' 20
periodic tables
    left-step 126, 128–31
    long-form 14, 17–18
    optimal 127–37
    short-form 13–14, 15
    versions 23–4, 122–6
        and triads 131–7
periods (rows) 12
Perrin, Jean 72, 75
phlogiston 30
phosphorescence 76–7
phosphorus (15), valency 40
physics and chemistry 28, 61–2
place names as element names 7
Planck, Max 85
planets as element names 6–7

platinum (78) and medicine 11
Platonic solids 1–2
plutonium (94) 113
polonium (84) 111–12
potassium (19)
    configuration 86, 90–1
    and medicine 11–12
    pair reversal with argon 71
    as part of a triad 36–7
predictions of new elements 63, 65–7
    failed 67–8
primary classification 24–6
promethium (61), name 6
protons and atomic number 22
Prout, William 35–6

Q

quantitative properties of
        elements 30
quantum mechanics 28, 98–102,
        106–8
    electrons as waves 104–5
    and the left-step periodic
        table 128–9, 130–1
quantum theory 27–8
    development 84–6
    quantum numbers 87–92, 106

R

radiation, early studies 84–5
radioactivity 73–4, 77
    to identify elements 5
Ramsey, William 69–70, 71
rare earths 17, 24
Rayleigh, Lord 69–70
reactivity of group 1 metals 15
relativistic effects 26–7, 118–21
repetition in elements 43
reversals
    of argon and potassium 71
    and isotopes 82–3
    of tellurium and iodine 64–5,
        70, 82–3

Index

rhodium (45), name 7
Richards, T. W. 82
Richter, Benjamin 31–2
roentgenium (111) 116
Röntgen, Wilhelm Konrad 73, 76
rubidium (37) and medicine 11–12
Rücker, Arthur William 70
Rutherford, Ernest 74–6, 77, 110
rutherfordium (104) 115, 118–19

S

s-block 129–30
Sacks, Oliver 6
salts 20
Saturn as atomic model 75
scandium (21), configuration 91
Schrödinger, Erwin 99, 104–5
scientists as element names 7–8
Seaborg, Glen 24–5, 113
seaborgium (106) 8, 120
secondary classification 24–6
Segrè, Emilio 76, 111–12
short-form periodic tables 13–14, 15
silicon (14), properties 16
Soddy, Frederick 81, 110
sodium (23)
    configuration 86
    with halogens 19
    as part of a triad 36–7
sodium chloride 61–2
Sommerfeld, Arnold 86–7
spectral lines 51–3
spectroscopy 3
stability, enhanced 116–17
Stoner, Edmund 88
Strömholm, D. 81
sublimation enthalpies 120–1
sulfur (16), valency 40
superconductivity 10–11
'superheavy' elements 26
super-triads 37–8
Svedberg, Theodor 81
symbols for elements 8–9

symmetry of periodic tables 128
synthesis 109–18
synthesized elements,
        behaviour 118–21

T

table of elements (not
        periodic) 39
technetium (43) 5
    synthesis 111
telluric screw 44
tellurium (52) and iodine 64–5,
        70, 82–3
thallium (204)
    classification 47
    name 7
thermonuclear test explosion
        114
Thomson, J. J. 73, 75, 92, 107
three-dimensional periodic
        tables 124–5
tin (50), properties 16
titanium (22), valency 40
transition metals 15–16, 24
    configuration 96
transmutation of elements 74,
        110–18
transuranium elements
    names 8
    synthesis 112–18
triads 36–41, 120
    alternatives to 61
    and versions of the periodic
        table 131–7
triple bonds 95
tungsten (74), name 9

U

univalent atoms 21
uranium (92)
    name 6–7
    and nuclear fission 113
    radioactivity 76–7

V

valency 16, 21–2, 40
variations of the periodic
    table 13–14
velocity of electrons 106
versions of the periodic table 23–4,
    122–6
  and triads 131–7

W

water 31–2, 33–5
waves 102–5
weight, equivalent and atomic 20–2
  see also atomic weight
Winkler, Clemens 66

X

'X's as isotopes 81
X-rays
  discovery 76
  frequencies of emissions
    79–80

Y

yttrium (39) and
    superconductivity 10–11

Z

Zapffe, Karl 50
zinc (30) and medicine 11

ONLINE CATALOGUE

Very Short Introductions

Our online catalogue is designed to make it easy to find your ideal Very Short Introduction. View the entire collection by subject area, watch author videos, read sample chapters, and download reading guides.